Natural Environment Research Council

INSTITUTE OF TERRESTRIAL ECOLOGY

PASTURE-WOODLANDS IN LOWLAND BRITAIN
A review of their importance for wildlife conservation

Paul T Harding

(ITE Monks Wood Experimental Station)

and

Francis Rose

(formerly Reader in Biogeography,
Kings College, University of London)

To
Philippa and Pauline

CONTENTS

ABSTRACT

The origins, form and composition of pasture-woodlands are summarized and a description is given of the flora and fauna particularly associated with this type of woodland. Evidence for considering the flora and fauna of these areas to be relics of those of the primeval forest, and to be indicators of continuity of habitat, is investigated. Particular attention is given to epiphytic lichens and saproxylic Coleoptera (beetles) and Diptera (flies). A selection of sites considered to be important for conserving the flora and fauna of pasture-woodlands is listed.

PREFACE

The following account of the conservation of epiphytes and saproxylic invertebrates in pasture-woodlands is a revised version of an earlier report to the Nature Conservancy Council (Rose & Harding 1978). Several years have elapsed since the submission of that report, during which time an essentially relevant book has been published on the subject of *Ancient woodland* (Rackham 1980). Frequent reference is made to this book in our account, particularly where the history of pasture-woodlands is concerned. Indeed, some sections of our account could be considered superfluous by anyone who has read Dr Rackham's scholarly work, but we hope that this biological account benefits from a brief historical review of the topic.

Current English usage in relation to wooded land can be a little confusing. The use of words with multiple meanings is frequent, if not prevalent, in common usage. Rackham (1980) provided a résumé of the names of types of wooded land in Britain and of the terms used in their management and description.

In the following account the names and terms used by Rackham are adhered to, except for the subject of our account, *pasture-woodlands.* Rackham derived the term *wood-pasture* from the medieval Latin term *silva pastilis* used, for example, in the Domesday Book. Our contention is that, whilst it is justifiable to refer to a wood/pasture system of management, it is grammatically more correct to refer to areas managed by this system as *pasture-woodlands.* Similarly, we refer to the other principal management types of woodlands as *coppice-woodlands* and *high-forest woodlands.* Our use of the term *pasture-woodland* differs in one respect from that of Rackham for *wood-pasture* (by including an additional type of pasture-woodland); our use of the term is otherwise synonymous with the term *wood-pasture* as used by Rackham (1976, 1980), Peterken (1977a,b, 1981), and others.

INTRODUCTION

Pasture-woodlands are areas of wooded land which have been used for the dual purpose of growing trees and grazing deer and livestock. Some relic areas of pasture-woodland provide an unique combination of continuity of habitat structure and site management, which has favoured the survival of species and communities dependent upon large old trees.

Areas we have considered to be, or to have been, pasture-woodlands were identified, first, by their present-day structure, primarily the presence of large old trees growing in, at least partial, open canopy, and, second, by the known history of the site and, where possible, its management. The survival of broadleaved trees into old age, whether by being left uncut or by deliberate management, such as pollarding, has been found to be a feature of areas managed as pasture-woodland. Many of the areas which were examined originated under a medieval wood/pasture management system, but are now incorporated into other types of woodland or, more commonly, into landscaped parks. Some are no longer grazed at all, except perhaps by wild deer.

The management of woodland in this way was an established practice in Anglo-Saxon times (Rackham 1980). It seems probable that it originated as a formalization of a grazing system in woodlands, especially those which were in the process of being cleared. It may, therefore, be as ancient a management system as coppicing.

Rackham (1977) has shown that the management of woodlands in Somerset, during the Neolithic period, was deliberate and included coppicing as an established practice. Although the area of pasture-woodland has been declining in Britain since the Middle Ages, we are convinced that this type of broadleaved woodland is now far better represented in lowland Britain than elsewhere in western Europe.

Relic areas of pasture-woodland, with old broadleaved trees, contain a structural variant not found in the other, more common, types of managed woodland, such as coppices and high forest. They are of ecological importance for the following 3 reasons.

1. The species of trees present, although undoubtedly influenced by centuries of management, including in some parkland sites exotic species planted for amenity reasons, may largely represent survivals of the genetic stock from the primeval forest, and ancient trees are often still present.

2. The epiphyte flora, in areas not seriously affected by atmospheric pollution, is often very rich and the communities of epiphytes are regarded as relics of similar communities present in the primeval forest at the relevant sites (Rose 1974, 1976; James *et al.* 1977).

3. Dead, dying and overmature trees often contain populations of local or rare saproxylic invertebrates - species which are dependent for some part of their life cycle on living, dying or dead wood, on associated fungi or Myxomycetes. These invertebrates, especially some beetles (Coleoptera) and flies (Diptera), are also regarded as relic populations, whose survival has been possible through the continuity of suitable habitat.

Until recently, only a few such areas had been studied in detail (for example, the New Forest in Hampshire and Windsor Forest in Berkshire) and even these were known only incompletely. Recent work by Rose (1974, 1976), Harding (1976, 1977a, 1978a-e) and Rose and Harding (1978) has suggested that many other areas are, or may be, of considerable importance for the conservation of wildlife associated with old trees. The desirability of conserving woodlands whose present structure has been determined by different types of management was emphasized by Peterken (1974, 1977a, 1981). One of the 5 types identified by Peterken was 'relics of medieval wood-pasture systems'.

Several pasture-woodlands were selected as being of high value for the conservation of wildlife in the *Nature conservation review* (Ratcliffe 1977); some were under the heading 'Mixed deciduous woodland: ancient parks and overmature woodland', others were more obliquely referred to under the vegetation types used to classify woodlands in that *Review*. It is clear that the conservation of woodlands, for wildlife, has been directed mainly towards those with rich and diverse assemblages of vascular plants, or towards presumed examples of vegetation types. However, following the preparation of the *Nature conservation review,* the Nature Conservancy Council commissioned a study of 'The fauna of the mature timber habitat' from the Institute of Terrestrial Ecology (Natural Environment Research Council). The study was to examine whether there were more sites which were of value for conserving the invertebrate fauna associated with old trees, especially in pasture-woodlands. This study was intended as a counterpart to surveys being made, independently, of the epiphytes in such areas (see section on Surveys).

ORIGINS OF PASTURE-WOODLANDS

Forests have grown in what are now the British Isles for many millions of years, but the primeval forests which preceded our present-day landscape developed only after the end of the last (Weichselian) glaciation. A long period (over 10 000 years) of arctic and tundra conditions with glaciers and ice-sheets came somewhat falteringly to an end about 10 000 years ago. At this point, the climate seems to have improved very rapidly, allowing the re-invasion of plants and animals from southern latitudes. The evidence of colonization of the post-glacial landscape by vegetation, leading to forest, was reviewed by West (1968), Pennington (1969) and Godwin (1975). Almost all lowland Britain was covered by forest 7000 years ago and the floristic composition of this forest, some 5000 years ago, before man had any significant impact on it, was reconstructed by Birks *et al.* (1975). (Rackham (1980) coined the term 'wildwood' for this primeval forest; this term is used henceforth to describe the unmodified forests of the last 10 000 years, the Flandrian stage.) Evidence summarized by Godwin (1975) suggests that during the Boreal (9500-7000 BP) the climate was at least as warm as the present climate and comparatively dry. It was followed by a wetter period, the Atlantic (7400-5000 BP), during which Palaeolithic and Mesolithic man spread over much of Britain. During the cool and dry conditions of the Sub-Boreal (5000-2700 BP), Neolithic and Bronze age man had a profound, if localized, effect on the wildwood. The development of, and changes in, the wildwood in Britain during the Flandrian stage have been reviewed by Rackham (1980).

Man clearly has been modifying the structure and reducing the area of the wildwood since the Neolithic period and these changes have inevitably had an effect on the local climate and upon the associated flora and fauna. The restriction of distribution, or loss of components, of the flora and fauna associated with the wildwood has been due mainly to the activities of man, doubtless supplemented by subtle changes in the global climate.

Management of the wildwood, probably to provide fodder for stock as well as underwood of selected dimensions, was almost certainly an established practice in the Neolithic period in Somerset (Coles & Orme 1976, 1977). The dual practices of clearance of the wildwood and management of at least part of the remainder inevitably continued to affect the flora and fauna. By the time of the Roman occupation of England (approx AD 40-400), it is clear that a substantial part of lowland Britain was no longer forested. During the Roman period, England was one of the major agricultural exporting areas of the classical world. The Anglo-Saxon colonization of England resulted initially in the continued use, for agriculture, of land cleared before or during the Roman period. An ever-increasing population led, inevitably, to further inroads being made into the remaining wildwood, to the extent that Rackham (1980) considered that possibly one third of England was covered by wooded land in the year AD 586, but that this figure had fallen to 15% by AD 1086. Woodland was clearly a carefully husbanded resource during the Anglo-Saxon period when the complexities of the multiple use of land, such as in the wood/pasture system, were carefully worked out (Rackham 1980).

TYPES OF PASTURE-WOODLAND

The survey of England summarized in the Domesday Book (AD 1086) gave an unique picture of the distribution of woodland and, to some extent, its use (Darby 1952-77; Rackham 1980). It makes a distinction, for some counties, between *silva minuta* and *silva pastilis* which Rackham (1976, 1980) interpreted as coppiced woodland and pastured woodland, respectively. Anglo-Saxon hunting areas are known to have existed (eg Harewood in Hampshire, Woodstock in Oxfordshire, and Clarendon in Wiltshire), but it is clear that pasture-woodlands were widespread and highly organized after the Norman Conquest (AD 1066). To some extent, this expansion of formal pasture-woodlands may have been a result of the introduction, by the Normans, of an additional *beast of the chase,* the fallow deer. However, it is certain that the legal and administrative systems adopted by the Normans, and their successors, also had an important influence.

The first 3, the main types of pasture-woodland dealt with below, are also described, with historical and geographic examples, by Rackham (1976, 1980).

FORESTS AND CHASES

Royal Forests and baronial Forests and Chases were administrative units of land, often very extensive in area, of which often only a small part was unenclosed land, which may not even have been woodland. Forests and Chases initially formed an administrative basis for government in the more remote areas and provided venison for the early medieval kings, their courts, and their wealthier subjects, at a time when fresh (and even salted) meat in winter was a highly valued commodity.

The importance of Forests, Chases and Parks for hunting by medieval kings and nobility, for sport, has probably been overstated in popular historical accounts. Those Forests and Chases that did not include woodland contained a nucleus of unenclosed land, the *physical* Forest, which could have been moorland, heath or fen. The remainder of the *legal* Forest included settlements, farmland and open pasture. Rackham (1980) recorded 142 Forests, of which about 80 were wooded, with an average of 5000 acres managed by the wood/pasture system. It seems that only 3% of England was covered by *physical* Forest, within the *legal* Forests, this at a time when Bazeley (1921) estimated that at least one third of the country was subject to Forest Law.

Forest Law came and went. With time, the Crown lost its interest in deer, better administrative systems developed, and the need for legal Forests diminished. Those Forests where the Crown owned the soil (eg Forest of Dean in Gloucestershire, and the New Forest) were managed by the Crown from the mid-17th century onwards in an attempt to grow timber, mainly for the Navy. Most of the wooded *physical* Forests remained as woodland, often as common land. Not until the Enclosure Acts did any serious reduction in area of the wooded component of former *legal* Forests take place. Woodland was taken into more rigorous management, or it was cleared for agricultural use. However, it is still a feature of many former Forests, particularly those which remained as *legal* Forests until late, that at least some of the natural element (the *physical* Forest) has remained. This is especially true of woodland, but many of these surviving woodlands are now managed as commercial forestry enterprises, and relics of the pasture-woodlands from former Forests are comparatively rare, although those that do remain are often extensive in area (eg Epping Forest in Essex, New Forest, and Windsor Forest).

PARKS

The term Park was used, in medieval times, to describe an enclosed area of land, usually, but not invariably, including some pasture-woodland intended for keeping deer.

Usually fallow deer were kept, but Rackham (1980) cited examples of parks for red and roe deer, white cattle, wild swine and hares; domestic stock seem also to have been kept in parks with deer. As with the Royal Forests, the prime importance of deer parks in medieval times was to supply fresh meat (venison), particularly during the winter months, as relief from the usual salted meat of domestic stock. Hunting for sport in parks, despite its romantic appeal, was probably not the main reason for their existence.

Although it is clear that many medieval parks contained woodland or trees in varying proportions, the typical 'parkland' combination of grassland, heath or bracken under scattered, often pollarded, trees was probably far from universal. The wide variety of forms and uses of parks recorded in medieval records were summarized by Rackham (1980). Most parks were privately owned, but villagers occasionally had rights to common grazing, to collect or cut wood, or were able to rent grazing (agistment).

Some 35 parks 'of woodland beasts' were listed in the Domesday Book and Rackham (1980) cited one example of an Anglo-Saxon deer park in Ongar in Essex, but there is no reason to believe that it was unique. The Normans, probably as a result of their conquest of Sicily, developed the Classical and Islamic traditions of emparking for their own purposes and imported, from the Levant and Near East, the fallow deer, a species already tried and tested in this role. The introduction of fallow deer, almost certainly after 1086, led to a rapid expansion in the number of deer parks, which continued into the 14th century. By the early 14th century, parks were a prominent feature of the landscape, and it is estimated by Rackham (1976) that there was the equivalent of one park in every 4 parishes (or 39 square kilometres). Cantor and Hatherley (1979) gave a figure of 1900 medieval parks in England, but these were rather irregularly distributed.

The origins of most medieval parks are obscure; at best, one has a record of a licence to empark, but this gives no idea of what sort of landscape was to be enclosed. Brandon (1963), from studies of parks in Sussex, gave evidence that many parks were formed on areas of 'waste' and probably contained relic areas of the wildwood which still existed in patches. However, Rackham (1976, 1980) believed that little, if any, unmodified wildwood remained into the 11th century. This latter view is reinforced by evidence from the Domesday Book that probably only 15% of England was wooded by 1086. What is certain is that, more often than not, medieval parks in lowland England contained at least some form of tree cover and were strongly correlated with the occurrence of woodland.

To confuse the issue, the establishment of parks has gone on, over the centuries. Some are obviously of post-medieval origin, others such as those at Ickworth and Sotterley in Suffolk embraced areas of agricultural land, with hedgerow trees and woodland, when formed in the late 17th or in the 18th century, to give the superficial appearance of a medieval park. The example quoted by Rackham (1980), of Thorndon Park in Essex, is of a park formed after 1774 to include Childerditch Common. However, this common was wooded, as a survey in 1774 recorded 2080 oak pollards and 1323 'hornbine' (hornbeam) trees. In many ways it is probable that the formation of most post-medieval parks was no different to that of earlier parks. The land-owner emparked the area that suited him best for his purposes, whether it was to keep deer for food, or to have a fashionably landscaped setting for his country house. Indeed, there are records of agricultural land being emparked as early as the 1290s, at Kings Langley in Hertfordshire and elsewhere (Cantor & Hatherley 1979). An interesting résumé of the uses of Woodstock Park in Oxfordshire, mainly during the 13th century, is given by Bond (1981). These uses include the extraction of timber, the maintenance of the herd of deer, pannage for pigs and the planting of pear trees in 1264. The landscaping of parks, popularly considered to be a prerogative of the 18th century, has a history going back at least as far as the 16th century. In fact, the 'naturalistic' landscaping of Brown, Repton and other 18th century landscape designers was to some extent a reaction against the more formal designs of previous centuries.

Although many medieval parks had only a short life, those that have survived have almost all undergone some modification at the hands of landscape designers. What we see at, for example, Blenheim Park in Oxfordshire is the much reduced remains of the royal Woodstock Park (it was 7 miles around in the 12th century), partially landscaped by Vanbrugh, heavily modified by 'Capability' Brown, now being replanted in places to replace the elms lost as a result of Dutch elm disease. The general history of parks after the medieval period is reviewed by Prince (1958), Hadfield (1967) and Rackham (1976).

In the 20th century, many parks have been reclaimed to arable land, or to virtually treeless pasture, but it is perhaps fortunate that many of those that have survived into this century are on what is potentially poor agricultural land and have so far escaped reclamation. The reclamation of parks, as with the destruction of country houses, seems to have taken place mainly before about 1960 (Gruffydd 1977). Of those that have survived, some are open to the public for recreation (eg Bradgate in Leicestershire and Richmond in London) or are attractive settings to stately homes, themselves open to the public (eg Longleat in Wiltshire and Knole in Kent). However, a surprising number of parks, particularly those that, as we shall show, are of value for wildlife conservation, are still privately owned, not open to the general public and, evidently, cherished by their owners despite the costs of upkeep.

WOODED COMMONS

Areas of land with common grazing-rights were plentiful in medieval England, and to some extent they overlapped with Royal Forests and even with some parks. Although many commons were open grassland or heath, some contained a woodland element and were therefore managed as pasture-woodland. Such use was recorded in Anglo-Saxon times, as at Felpham in Sussex, where swine pastures on a pasture-woodland common called Palinga Schittas may be referable to the modern wooded common, The Mens (Tittensor 1978).

Pollarding seems frequently to have been the main method of managing trees on wooded commons. Areas with pollards remain in a few localities, and in some cases (eg Burnham Beeches in Buckinghamshire and Ebernoe Common and The Mens in Sussex) the commons are almost entirely wooded. Few ancient, lowland, wooded commons survive now outside the Weald, the Chilterns, in Dorset and in Wiltshire.

Many commons that were open grassland or heath have become wooded in the last 2 decades, largely as a result of the post-myxomatosis decline in rabbit populations and the decline in the use of common grazing rights. These commons contain only comparatively young trees, with no continuity of age classes, and, as a result, are of little interest for epiphytes or saproxylic invertebrates, although they may be of outstanding interest for other flora and fauna.

Many of the surviving wooded commons are now used mainly for public recreation. Grazing of domestic stock is rare and the cutting of wood from pollards ceased in most areas over 50 years ago. Several wooded commons are protected under the Commons Registration Act of 1965 (Denman *et al.* 1967).

WINTER-GRAZED WOODLANDS

The enclosure of domestic stock, or the natural descent of sheep and deer from upland summer pastures into adjacent valley or lowland woodlands in winter, is common in parts of western and northern Britain. In some areas of the west highlands of Scotland, this form of local migration is essential to the economy of sheep farming, as it is to the survival of red deer herds (Mitchell *et al.* 1977).

Most of the woods used for winter grazing in upland areas are now derelict coppice woodland and the incidence of winter grazing with the management of the woodland to retain old trees is rare. A few areas are essentially of this type (see Table 6), but probably the most important example is the Horner Coombe complex of valley woods on the north-east side of Exmoor, which has traditionally been one of the main wintering grounds for deer from the moor.

Our definition of winter-grazed woodlands may not fit comfortably with the categories of Rackham (1976, 1980) and Peterken (1974, 1977a,b), but, in our opinion, in the cases of epiphytes and probably of saproxylic invertebrates, they are sufficiently distinct to merit consideration as a further type of woodland managed under a wood/pasture system.

SURVEYS

A study of the number of species of epiphytes, their density and abundance in selected woodland areas throughout Britain was begun in 1969 (Rose *et al.* 1970). Some of the results of this study have already been published (Rose 1974; Rose & Wallace 1974; Hawksworth *et al.* 1974; Rose & James 1974). More detailed results, with special reference to the use of lichens as indicators of age and environmental continuity in woodlands, were given by Rose (1976) and James *et al.* (1977). Early results from this work were used by Ratcliffe (1977) in selecting areas considered to be of value for the conservation of epiphytes. Numerous reports on lichens in woodlands throughout Britain, including many pasture-woodlands, have been prepared for the Nature Conservancy Council by one of us (FR), sometimes in collaboration with other lichenologists. These reports (Appendix 1) have not been published and most of them are deposited at the appropriate regional offices of the Nature Conservancy Council.

In 1982, the British Lichen Society produced an assessment for the Nature Conservancy Council of 1700 woodland sites throughout Britain, including some pasture-woodlands (Fletcher *et al.* 1982).

An inventory was compiled of sites in lowland Britain considered to be of possible, potential, or known value for the conservation of invertebrates associated with mature and overmature deciduous trees and dead wood (Harding 1976, 1978e). The assessment of sites was based, wherever possible, on knowledge of the invertebrate faunas from all available sources, also on the structure of the woodlands and on historical records. The final inventory (Harding 1978e) included over 400 sites, of which approximately 100 were visited between 1975 and 1978 to assess their present-day appearance and management.

Description of the sites visited were given in Harding (1976, 1977a,b, 1978a-d, 1979a-c, 1980, 1981). Details of the attributes recorded during visits were given by Harding (1981).

The effects of air pollution on epiphytes were discussed by various authors in Ferry *et al.* (1973). It should be noted that relatively few pasture-woodland areas known to be in the more heavily polluted zones described by Hawksworth and Rose (1970, 1976) have been studied for epiphytes. Those that have been studied now have poor epifloras (Table 1).

The effects of air pollution on the distribution of woodland invertebrates, other than those species which feed on epiphytes (Gilbert 1971), are largely unknown. Several of the sites in heavily polluted areas (Table 1) are known to be of importance for the conservation of saproxylic invertebrates.

Table 1. The number of taxa of epiphytic lichens recorded from 11 sites known to be in polluted zones. (Figures in brackets are the number of taxa recorded during the 19th century).

	Number of epiphytic lichen taxa
Woburn Park, Bedfordshire	51
Chatsworth Park, Derbyshire	16
Epping Forest, Essex*	38 (118)
Hatfield Forest, Essex	63
Brocket Park, Hertfordshire	45
Cobham Park, Kent*	39
Gopsall Park, Leicestershire	12 (106)
Grimsthorpe Park, Lincolnshire*	29
Richmond Park, London*	8
Sherwood Forest, Nottinghamshire*	8
Sutton Park, West Midlands	48

*Sites known to be of importance for saproxylic invertebrates.

These surveys of a wide range of woodlands in Britain have demonstrated that, in lowland Britain, some areas of extant or former pasture-woodland containing old trees support epiphytes and invertebrates not, or little, recorded in other types of lowland woodland, such as coppice-woodlands or planted high forest, or in other habitat types (Rose & Harding 1978). However, many of these local species of epiphytes do occur also in apparently relic woodlands on steep escarpments or in ravines in western and northern areas of Britain, such as the remoter valleys and scarps, at relatively low altitudes, in Wales, Lakeland and western Scotland. The same appears to be true for a limited number of saproxylic invertebrates.

PRESENT-DAY STRUCTURE

Aspects of the structure of some pasture-woodland areas have been studied by various authors (Table 2), but, as noted earlier, no single description of pasture-woodlands is adequate, a view reinforced by Rackham (1980). However, a synthesis of the accounts listed in Table 2 would characterize almost all such areas that remain today.

Table 2. Recent studies of some pasture-woodlands.

Sites	Authors
Windsor Park and Forest, Berkshire	Welch (1972)
Shute Deer Park, Devon	Harding (1977, 1980)
Epping Forest, Essex	Corke (1978), Hanson (1983)
New Forest, Hampshire	Tubbs (1968), Rose & James (1974)
Brampton Bryan Park, Hereford/Worcester	Harding (1979a)
Moccas Park, Hereford/Worcester	Harding (1977b)
Wyre Forest, Hereford/Worcester	Hickin (1971), Hawksworth & Rose (1969)
Lullingstone Park, Kent	Pitt (1984)
Dunham Massey Park, Manchester	Smith (1978), Harding (1979c)
Nettlecombe Park, Somerset	Rose & Wolseley (1984)
Staverton Park, Suffolk	Peterken (1969)
The Mens, West Sussex	Tittensor & Tittensor (1977), Tittensor (1978)
Dalkeith Old Oakwood, Lothian	Fairbairn (1972)
Dynevor Deer Park, Dyfed	Harding (1979b, 1981)

TREES

The tree component in pasture-woodland varies considerably in density, species structure and management. Three main types of management of trees can be recognized - pollards, maidens and coppice. The form, origin and uses of these types are described and illustrated by Rackham (1976, 1980).

Pollards are trees which have been cut at 2 -3 m above ground, out of the reach of browsing animals, and allowed to grow again from the bolling (stump) to produce crops of shoots which were periodically cut to provide wood for fuel, building material and, in the case of oak, bark for tanning. In parks and forests, wood was cut to provide browse-wood for deer, especially in hard winters, just as holly is still cut in the New Forest (Tubbs 1968) (see also Spray 1981).

Pollarding became illegal in the New Forest in 1698 (Tubbs 1968) because of the need for straight, long timber for shipbuilding. The practice probably began to decline at about the same time elsewhere. Edlin (1971) stated that, with few exceptions, it ceased as a practice in England in about 1870, but to this day trees in a few localities are pollarded as a commercial practice (eg Mersham Hatch Park in Kent).

Coppicing is the management of trees and shrubs to produce multiple small growths from a single root stock by repeated cutting at, or near, ground level. Because deer and stock would continually remove the young shoots from coppice stools, it would clearly be difficult to run deer, or stock, in coppiced areas. This problem was resolved by the use of temporary enclosures around young coppice growth until it reached a size where browsing could not cause serious damage. Thus, it was possible for some pasture-woodlands to have few if any old trees because all the available woodland was coppiced. Some coppices have, however, been neglected since the 16th century and, especially in parts of the New Forest, they are now areas of fine, uneven-aged high forest. However, some early 'encoppicements' in the New Forest seem to have been enclosed areas of forest used for growing timber trees, as well as underwood, rather than coppice in the accepted sense (Flower 1977).

Maidens. Maiden trees are of seedling origin and have been neither pollarded nor coppiced. They are a common and important feature of some areas.

These types of management are not as simple as the above account might suggest. Pollards, or coppice, may have been cut once or many times, giving rise to a variety of growth forms. In some areas, former coppices, particularly of oak and ash, have been converted to high forest by the selection (singling) of one growth from each coppice stool to form a standard tree.

Trees may be damaged at an early stage of their development by, for example, browsing animals, frost or drought, giving rise to curious growth forms. The competition of neighbouring trees for light also affects growth form, leading to tall straight stems where a tree is growing in an even-aged stand, or to a wide and spreading crown where the tree is growing without competition.

FIELD LAYER PLANTS

The vascular plants of pasture-woodlands are usually poor in species for 2 main reasons: (1) the remaining areas are mainly on soils with a low fertility, particularly sands, gravels and neutral or acidic clays; (2) grazing eliminates many typical woodland plants by repeated physical damage. It is also possible that the continued removal of minerals, particularly calcium phosphate, from the nutrient cycle by deer and stock that finished up on the dining table may have a long-term effect on the vegetation, particularly in deer parks.

In southern England, there are 2 plant species which, although never common, are characteristic of old woodlands of all types and persist in pasture-woodland. These are butcher's broom (*Ruscus aculeatus*), a spiny, low shrub, and wood spurge (*Euphorbia amygdaloides*) which contains an acrid, unpalatable latex.

Many pasture-woodlands on sands and gravels have a ground flora of bracken (*Pteridium aquilinum*), heather (*Calluna vulgaris*), bilberry (*Vaccinium myrtillus*), honeysuckle (*Lonicera periclymenum*), grasses (*Agrostis* spp., *Holcus* spp.) and mosses. Similar soils under oak woods which are, and have been, little grazed, in western and northern Britain carry woodrush (*Luzula sylvatica*) as an abundant field layer species, but this is absent in most pasture-woodlands.

Few sites in southern lowland Britain combine a rich epiphyte flora with a rich ground flora, other than in streamside alluvial flood plains and in carrs (eg parts of the New Forest, such as Wood Crates). The conditions for the survival of both floras rarely occur together, although a few long enclosed ancient woods on basic clay soils (eg High and Low Woods at Longleat) do combine both features. Similarly the formerly unenclosed and grazed coppice-woodland and high forest areas of Cranborne Chase in Wiltshire have rich epiphyte and field layer floras. Other examples of areas combining richness in both types of flora are Roydon Woods near Brockenhurst in Hampshire, Langley Woods near Hamptworth, and Great Ridge Wood in Wiltshire. Roydon Woods are now enclosed woodland, but part contains large old trees, perhaps dating from when this area was within the New Forest. Langley Woods and Great Ridge Wood are also now enclosed woodlands, but both contain areas whose structure (and bank systems) suggests that parts were once pasture-woodland.

TREE SPECIES AND THEIR ASSOCIATED EPIPHYTES AND SAPROXYLIC INVERTEBRATES

NATIVE SPECIES OF TREES

In most pasture-woodlands 2 species, pedunculate oak (*Quercus robur*) and beech (*Fagus sylvatica*), are the overwhelmingly dominant trees today. Other species are less common, although they may dominate locally, but some are of considerable importance for the special habitats they provide for epiphytes and invertebrates. In the following account, oak and beech are considered first. The remaining species are dealt with in systematic order.

Pedunculate oak (*Quercus robur*)

The distributional and taxonomic history of both native species of oak were reviewed by Gardiner (1974), and Rackham (1980) discussed the history of oak woodland in Britain. Pedunculate oak is now the more widespread of the 2 species, partly due to selective preference for the timber of this species over the last 2 centuries, but also because of its greater ability as a pioneer species. In most parks, and in wooded commons on heavier soils, pedunculate oak is by far the commoner of the 2 native oaks.

Sessile oak (*Q. petraea*)

This is seldom as common as pedunculate oak in pasture-woodlands, although in a few places (eg parts of the New Forest) it does predominate. Many old parks (eg Moccas Deer Park in Hereford/Worcester) have small numbers of this species and it also occurs on some wooded commons on lighter soils.

In his paper on the epiphytes of oak, Rose (1974) compared the numbers of lichen taxa recorded as epiphytes on 13 tree genera of species in the British Isles, and demonstrated the importance of the 2 native species of oak. Revised totals, based on current records from Britain, are given for 15 genera or species in Table 3. A total of 326 taxa is recorded from oak, the next highest totals being for ash (*Fraxinus excelsior*) and beech. Many epiphytes of oak take a long time to develop and either occur only on mature or ancient trees, or are found to be well developed only on such trees, eg *Lecanactis premnea* and *L. lyncea*.

A comparison of the numbers of phytophagous insects associated with various species of tree in Britain was given by Southwood (1961). Revised totals for phytophagous insects and mites associated with 28 British tree and shrub species were given by Kennedy and Southwood (1984). In the latter paper, the 2 native species of oak total 423 species of associated fauna, with only *Salix* (5 species) scoring more species (450). Total figures for 15 species or genera of trees and shrubs, taken from Kennedy and Southwood (1984), have been included in Table 3.

These data justify the common belief that oak woodland, particularly ancient woodland containing old trees and dead wood, is of great importance for invertebrates, a point emphasized by Morris (1974).

It is interesting that only very few, if any, lichens are strictly confined to oak (Rose 1974). Unfortunately, no list of the saproxylic invertebrates of oak has ever been compiled, but many of our rarest Coleoptera are associated, as larvae, with oak. Notable species such as *Plectophloeus nitidus, Lacon querceus, Pyrrhidium sanguineum, Hypebaeus flavipes, Trixagus brevicolis* and several species of *Ampedus* have been recorded only from oak in Britain, and it seems probable that oak is the most important tree for saproxylic Coleoptera.

Beech (*Fagus sylvatica*)

The popular fallacy that beech is a calcicole species is refuted by its common occurrence in many pasture-woodland areas on acidic soils. It is an aggressive species in the present oceanic climatic conditions and tends to invade open oak wood on well-drained soils. In the New Forest, and probably elsewhere, it is more abundant than formerly because of the climatic conditions, which have permitted it to recolonize after earlier felling of oak (Flower 1980; K Barber, pers. comm.). It is certainly native in the New Forest and in other parts of southern England and south Wales, possibly as far west as Cornwall. Its northern limits of natural distribution are difficult to define. From pollen analyses, Oxford (1975) confirmed that beech was present in Epping Forest in about 2320 BC. It may well be native elsewhere in East Anglia (eg Felbrigg in Norfolk) and as far north as Lancashire (Godwin 1975). It has been planted widely throughout Britain from the 17th century onwards. It is commonly found as a planted tree in pasture-woodland (eg Savernake Forest and Longleat Park in Wiltshire, and Knole Park, Kent), and was an essential component of much 18th century landscape planting.

Beech is of considerable importance for epiphytes. A total of 213 lichen taxa has been recorded from it in Britain. Despite its smooth bark it carries an epiflora, in the New Forest, very like that of oak (Rose & James 1974), though with a few species almost or quite confined to it (eg *Parmelia minarum* and *Catinaria laureri*). Beech has few epiphytes in chalk scarp woodlands, probably because the bark dries too rapidly after rain in these well-drained, often exposed, sites on very porous soils. In common with oak, ash and elm, beech has many saproxylic species of Coleoptera and Diptera, several of which are very rare and localized. In the New Forest and Windsor Forest, beech rivals oak for the occurrence of the rarer saproxylic invertebrates. Species of Coleoptera, such as *Eucnemis capucina, Rhyncolus truncorum* and *Limoniscus violaceus,* and the fly *Chrysopilus laetus* are known to breed only in beech in Britain. It seems probable that more species of Diptera are associated with the dead wood of beech than with that of oak.

Table 3. Total numbers of taxa of epiphytic lichens and phytophagous insects and mites associated with trees and shrubs in Britain.

Tree/shrub		Numbers of lichen taxa	Numbers of insect and mite species (†)
Field maple	*(Acer campestre)*	101	51
Sycamore	*(Acer pseudoplatanus)*	194	43
Alder	*(Alnus glutinosa)*	116	141
Birch	*(Betula pubescens/pendula)*	134	334
Hornbeam	*(Carpinus betulus)*	44	51
Hazel	*(Corylus avellana)*	162	106
Beech	*(Fagus sylvatica)*	213	98
Ash	*(Fraxinus excelsior)*	265	68
Holly	*(Ilex aquifolium)*	96	10
Pine	*(Pinus sylvestris)*	133*	172
Oak	*(Quercus petraea/robur)*	326	423
Willows	*(Salix* spp.)	160	450
Rowan	*(Sorbus aucuparia)*	125**	58
Limes	*(Tilia* spp.)	83	57
Elms	*(Ulmus* spp.)	200	124

(†) Numbers from Kennedy & Southwood (1984) Table 1.
* In native Caldeonian pinewoods.
** Mainly in upland woodlands.

Yew (*Taxus baccata*)

With its very thin, peeling bark, yew rarely carries any epiphytes of importance in pasture-woodland, though ancient specimens in some churchyards in extreme southern England sometimes have the lichen *Opegrapha prosodea,* which is otherwise very rare and largely confined to very old oaks in a few southern parks and forests. Yew is frequent in old pasture-woodlands on a variety of soils throughout south and west England, but is strangely absent from them in East Anglia. The fauna of yew is similarly poor in species and no Coleoptera or Diptera are known to be restricted to its wood, although a few common bark-dwelling Coleoptera have been recorded from it.

Field maple (*Acer campestre*)

Field maple is a widespread native species occurring particularly on the more base-rich soils. Large standard trees are a feature of certain areas such as Wychwood Forest in Oxfordshire, Cranborne Chase in Wiltshire, Grimsthorpe Park in Lincolnshire, and, more locally, in the New Forest and Eridge Park in Sussex. It has a most favourable bark for epiphytes; the texture is porous and corrugated, the pH is high (5.5-6.5) and it has a high base-exchange capacity. Many very interesting lichen and bryophyte epiphytes occur on this tree in pasture-woodlands of southern England, but the total epiflora is small (Table 3), probably because there are relatively few large specimens in pasture-woodlands. Hence, compared with other tree species, it is unable to contribute much to the total surface available for epiphyte colonization. The fauna associated with its wood is poorly recorded. It supports species in common with oak and beech, but at Windsor Forest the importance of field maple lies in the occurrence of the beetle *Gastrallus immarginatus* (Allen 1956). The flowers of this species are attractive to a few Coleoptera, notably *Osphya bipunctata,* whose larvae may also bore into its wood.

Blackthorn (*Prunus spinosa*)

Areas of blackthorn scrub in damp sheltered situations in the New Forest are sometimes rich in *Usnea* spp., but otherwise this species is of little importance for epiphytes in the types of habitat present in pasture-woodlands. A few common Coleoptera have been recorded as being associated with the dead wood of blackthorn and the flowers are an important nectar source.

Hawthorns (*Crataegus* spp.)

C. monogyna (and very rarely *C. laevigata*) may be common, often as ancient specimens, in pasture-woodlands. The former is particularly characteristic of some old deer parks. Neither species is of much importance for epiphytes. The wood of hawthorn is utilized by a number of fairly common species of Coleoptera, but it is the blossom of the shrub that is particularly important in pasture-woodland. The flattish corymbs of flowers, rich in nectar, are attractive to many insects and the emergence of adults of many species of Coleoptera and Diptera, associated with wood as larvae, coincides with its flowering. In some pasture-woodlands, hawthorns provide almost the only source of nutrition for nectar-feeding adult insects. Other important sources are the flowers of blackthorn, rowan, elder and of some Umbelliferae.

Rowan (*Sorbus aucuparia*) and other *Sorbus* species

Rowan is widespread as a native species on acid soils in British woodlands; it carries very important epiphyte floras in the Scottish highlands, but poor ones in lowland Britain. The other *Sorbus* species play a very minor role as phorophytes in Britain. Although wild service (*S. torminalis*) appears to be a good indicator of ancient woodland sites, and is a feature particularly of several Wealden pasture-woodlands, it is rarely numerous and appears to have an unfavourable bark for most epiphytes. The wood of *Sorbus* spp., because trees rarely reach any great size, seems not to have any uncommon fauna recorded from it. The blossom of rowan, like that of hawthorn, is of some importance for nectar-feeding insects.

Limes (*Tilia* spp.)

Although limes were clearly important in the primeval forests, as has been demonstrated by many workers on fossil pollen records (see Godwin 1975), they are now rare species in old pasture-woodland, except as planted trees, usually the common lime (*T.* x *vulgaris*). Lime woods occur locally and lime trees are still a feature of some little grazed escarpment, ravine or gorge woodlands. A few specimens occur on scarps in some parks in Herefordshire. The epiflora is not rich probably because of the rather hard fibrous bark which does not hold water. The entomology of the wood of lime is little worked, but at least one beetle, *Ernoporus caucasicus*, is known only from lime in Britain (Cooter 1980; Harding 1982). Once dead, lime bark degrades rapidly to a fibrous mass; therefore, it is not available to many bark-dwelling species. Once the bark has decayed, the wood desiccates rapidly, and it is little used by saproxylic species.

Elms (*Ulmus* spp.)

Wych elm (*U. glabra*) occurs quite commonly as ancient pollards or coppice stools, especially in ravines or on escarpments in some old parks, particularly in western Britain. It is an important phorophyte, but its existence in Britain is threatened by the spread of the new virulent strain of Dutch elm disease (*Ceratocystis ulmi*). This same disease also threatens the other, usually planted, species of elm which are an important constituent of parks landscaped in the 17th to early 19th centuries. These carry, in open, well-lit situations, the lichen community known as *Xanthorion* (see James *et al.* 1977). Many invertebrates favour the decaying wood of elm for breeding sites. A notable species is the rare beetle *Elater ferrugineus,* larvae and pupae of which have been found commonly in the rotten interior of elms in Windsor Park, felled because they had died of Dutch elm disease (P Cook, pers. comm.). The spread of Dutch elm disease has also led to an increase in the occurrence of another beetle *Aulonium trisulcum,* a predator of elm bark beetles (*Scolytus* spp.) in elm bark (Marshall 1978).

Birches (*Betula pubescens* and *B. pendula*)

The 2 tree-birches are very widespread in pasture-woodlands, often as old relic trees from phases of regeneration due to reduced intensity of grazing in the past. Both birches appear to be important colonists on a wide range of soil types in such sites. Birches have smooth, very acidic bark when young, and hence are not particularly rich in epiphytes in lowland Britain, although old specimens, with more rugged bark, may support a number of the more calcifuge lichens. In the high rainfall areas of northern and western Britain, however, they bear a richer, more specialized epiphyte flora, including the lichen association *Parmelietum laevigatae* and many bryophytes. A considerable number of invertebrates is associated with birch wood, but only a few species appear to be limited to it. However, in north-west England and in Scotland several local species have been recorded mainly from birch. The birch bracket-fungus (*Piptoporus betulinus*) occurs commonly on dead and dying birch trees and supports a variety of Coleoptera.

Alder (*Alnus glutinosa*)

Alder is widespread as a native of moist pasture-woodlands in Britain, especially along streams, on spring lines, and in areas of carr. Generally in the lowlands its epiphyte flora is extremely poor, probably because of its hard acid bark (pH usually below 5.0) and the poor water retention of the bark surface. However, in north-west England and in the Scottish highlands it may carry several interesting lichens. Its wood is soft and decays rapidly. Consequently, it is probably of little importance for the fauna associated with dead or decaying wood. It is, however, utilized by several Diptera, a few common Coleoptera, and the alder woodwasp (*Xiphydria camelus*).

Hornbeam (*Carpinus betulus*)

Hornbeam is apparently a native species only in south-east England and East Anglia, and occurs as far west as the Bristol area only in small numbers. It is, however, important locally, often as old pollards in pasture-woodland areas in parks in Kent and in Epping and Hatfield Forest, Essex. Today, much of its range suffers from too much air pollution for it to support many epiphytes, but in parts of east Kent, with a less polluted atmosphere, it has a flora of crustose lichens rather like that of beech - presumably because of its rather similar, smooth bark texture. The fauna associated with hornbeam, especially as ancient pollards, has similarities with that of oak and beech. Where hornbeam occurs, it is frequently an important contributor to the dead wood habitats.

Hazel (*Corylus avellana*)

Hazel is an important phorophyte for lichens, especially those of smooth bark in higher rainfall areas which are relatively free of air pollution, and where ancient specimens occur. Except in a few sites, as at Cranborne Chase, and Wychwood Forest in Oxfordshire, it is generally very rare in pasture-woodlands and so plays only a minor role. Hazel is not a particularly important species for fauna associated with wood, but in neglected coppice woodland several uncommon Coleoptera have been recorded breeding in large dead stems.

Holly (*Ilex aquifolium*)

Holly occurs abundantly in some old pasture-woodlands on acidic soils, notably parts of the New Forest, The Mens and Ebernoe Common in Sussex and in some East Anglian sites, such as Staverton Park in Suffolk and Merton Oaks in Norfolk. Holly in pasture-woodland was used as an important browse wood for stock and deer in hard winter weather. An interesting variant of this use is found in Shropshire, Herefordshire, Yorkshire and elsewhere, near upland areas, where holly-dominated pasture-woodlands occur, often intermixed with hawthorn 'orchards'. These areas were probably used as winter pastures for hill sheep, the foliage being cut in controlled quantities from the holly pollards (Radley 1961; Spray 1981).

Holly has a limited, but specialized, lichen flora with some very local crustose species. The fauna of holly is not particularly rich, but is distinctive. However, few invertebrates appear to be restricted to its wood.

Willows (*Salix* spp.)

These, especially *S. atrocinerea,* may be very important phorophytes for epiphytic lichens in areas of carr in pasture-woodlands in highland Britain, with rare and specialized species such as *Heterodermia obscurata* and *Cetrelia olivetorum* occurring on it as far south as Devon. In the east of Britain, few species of importance occur on willows. Some of the valley carrs of the New Forest, however, have interesting epiphytic lichens (*C. olivetorum, Parmelia laevigata*) that are otherwise oceanic and western in distribution. Willows seldom reach a size, or condition, in pasture-woodland to be important for invertebrates associated with the wood, although some larger waterside trees are inhabited by several uncommon Diptera and Coleoptera.

Ash (*Fraxinus excelsior*)

Ash is frequent in many pasture-woodlands on the more base-rich soils of streamside alluvial flood plains, and often as a planted tree in parks. It is probably native throughout lowland Britain. The bark of ash is fissured and rather similar to that of oak, but often of higher pH. It supports a rich lichen epiflora - 265 taxa are now recorded (Table 3). Decaying ash provides conditions for invertebrates similar to those found in beech. Ash often supports a rich fauna, but it is doubtful whether many Coleoptera or Diptera breed exclusively in it in Britain.

INTRODUCED SPECIES OF TREES

Some introduced species are sufficiently common in pasture-woodlands, especially parks, that their role as phorophytes, and as habitat for some saproxylic invertebrates, deserves mention.

Sycamore (*Acer pseudoplatanus*)

Sycamore is the most important introduced tree in Britain as a phorophyte for lichens and bryophytes. Introduced in the 15th or 16th century, it is so much at home in our climate that it has become extensively naturalized in many woodlands on slightly acid to neutral soils, and regenerates freely. Sycamore has been extensively planted in parklands, and in open areas tends to carry a similar epiflora to elm, with the *Xanthorion* community often well developed; its bark is of high pH and of high base-exchange capacity. It is very rare in the New Forest (except Matley Wood), but in a few very old parks, such as at Melbury and Lulworth in Dorset, old specimens occur in ancient oak woods with more or less closed canopy, and occasionally carry the Lobarion community of 'old forest indicator' lichens.

Walnut (*Juglans regia*)

Walnut is often found as old planted specimens in open parkland. It sometimes carries a rich epiflora of the *Xanthorion* type, and is particularly important in East Anglia, as at Sotterley Park where the lichen *Normandina pulchella* (now only known in one other East Anglian site) occurs on it.

Sweet chestnut (*Castanea sativa*)

Sweet chestnut is an ancient introduction that is often found as old specimens in parkland. Perhaps because of its fibrous bark of poor water retention, it generally bears a poor epiphyte flora in the British Isles, but in southern Europe (eg Cevennes, Provence and Tuscany) rich *Lobarion* communities occur on it in montane forests at 400-1000 m altitude.

Horse chestnut (*Aesculus hippocastanum*)

Horse chestnut is plentiful in some parklands as a result of plantings. It sometimes bears epiphyte floras of the *Xanthorion* type, but the heavy shade that this tree usually casts upon its own trunk tends to preclude the development of rich epiphyte communities, and it is unimportant in this respect. Few uncommon Coleoptera or Diptera have been recorded from alien tree species, except those particularly associated with the fruiting bodies of fungi on decaying wood; several of the species recorded, for example by Crowson and Hunter (1964), were from horse chestnut.

It has been suggested (F A Hunter, pers. comm.) that several species of Coleoptera, which are typically associated with oak, are able to utilize suitably aged and decayed sweet chestnut. The same may be true with some exotic species of oak (R C Welch, pers. comm.). Duffy (1952) stated that, in the case of the larvae of long-horn beetles (Cerambycidae), the physical condition of dead wood was often more important than the species of tree to which it belonged.

THE HABITATS OF EPIPHYTES AND INVERTEBRATES

WOODLAND STRUCTURE

Many groups of organism associated with woodland show greater species diversity and biomass in ecotones, such as the edges of gaps, glades, rides and lawns in woodland. These ecotones all have features in common - sheltered conditions (leading to high air humidity and/or low evaporation) and adequate illumination. The concentration of species and numbers of organisms (epiphytes, adult insects, flowering plants and birds) in such ecotones within woodland areas is so marked that it leads one to conclude that the wildwood may have had numerous glades and gaps, and was not uniformly dense. Several epiphytic lichens occur in dense high forest stands, but their numbers are normally relatively small, compared with those of the better-lit edges, and in such stands many species are confined to branches in the crown where illumination is greater. The invertebrates of the canopy are poorly studied in temperate areas, but from studies in tropical forests it is clear that many invertebrates, especially insects, live out most or all of their life cycle in the canopy.

Gaps in the canopy of the wildwood would have existed due to several causes, principal of which are:

1. the presence of areas of open water or wetlands;
2. the natural death and decay of trees;
3. the activities of large herbivores and man.

It is inevitable, especially with the relatively high rainfall of the Atlantic period (7400-5000 BP), that the wildwood was interspersed with rivers, lakes and wetlands. Few examples of forest wetlands remain in Britain, but possible examples are some of the bogs in the New Forest. Ancient wetlands of this type retain elements of the flora which existed before the Atlantic forests developed (Pigott & Walters 1954; Rose 1957).

Gaps would have been created by the natural, but usually gradual, death and decay of overmature trees (Streeter 1974), but these gaps were probably short-lived as regeneration took place.

Evidence of the presence of large herbivores in the wildwood was reviewed by Grigson (1978). She concluded that, apart from red and roe deer and wild swine, all of which survived into the times of documented history, aurochs (*Bos primigenius*) and probably a species of horse (*Equus* sp.) were present during the Atlantic period. The herbivores would have created and maintained, by grazing and browsing, gaps created initially by the death of old trees, and, by trampling, trackways to sources of water. A woodland structure of this type can be seen in many of the 'Ancient and Ornamental Woodlands' of the New Forest today, and in the wooded areas of some ancient parks, such as Boconnoc in Cornwall, Melbury in Dorset and Parham in Sussex, that are still grazed.

The role of man in modifying the wildwood has tended to be underestimated by earlier authors. However, several recent papers (for example, Smith 1970; Jacobi 1978; Coy 1982) have emphasized the considerable modifications to the wildwood wrought by Mesolithic man. It is quite clear that, in the Somerset Levels, early Neolithic man had developed methods of using the forest in an organized way (Rackham 1977). Neolithic man not only had access to the wild animals of the forest as a food resource, but also had domestic cattle, sheep and pigs, all of which would have had their effect on the extent and structure of the wildwood.

If such open canopy areas were not a feature of the wildwood, it is surprising that such a large proportion of the flora and fauna of our present woodlands should be species adapted to well-lit conditions. In closed-canopy woodlands, that are not regularly coppiced, well-lit and humid conditions are rare and ill developed, and must always have been so if there were no open areas of the types described above.

There is much available evidence that evolution of new species is a slow process, compared with the human time scale, and it is thus likely that the species assemblages of European broadleaved woodlands have originated over a long period of time (see West 1968; Coope 1970) stretching back beyond the Interglacials into the later Pliocene period at least.

It can therefore be argued that the mosaic and partly open structure of pasture-woodlands today is probably far closer to that of much of the wildwood than is that of the densely planted high forest areas of today, or even that of old naturally developed woodlands which, through lack of any grazing, etc, have become very dense and without any glades.

EPIPHYTE HABITATS (see Rose 1974; James *et al.* 1977)

Illumination. Well-illuminated tree trunks in open parkland carry epiphytes adapted to good illumination and to frequently intense drying-out by sun and wind. The epiphytes that occur in such situations include some also found in closed woodland, but for the most part they are taxa equally characteristic of roadside, hedgerow or pasture trees, such as *Parmelia caperata, Evernia prunastri* and *Pertusaria amara.* In such conditions, where there is enrichment of the bark by animal urine and faeces, mostly carried as dust by wind on to the tree boles, the characteristic community of lichens known as the *Xanthorion* tends to develop, particularly on rough bark of high water retentivity and base-exchange capacity. This community contains orange-yellow foliose lichens, such as *Xanthoria parietina* (which gives the community its name), grey or white foliose lichens of the genera *Physcia, Physconia* and *Phaeophyscia,* sometimes *Anaptychia ciliaris,* and various other foliose, shrubby and crustose taxa, together with certain bryophytes such as *Camptothecium sericeum, Orthotrichum* spp., and in the extreme south of England *Leptodon smithii.* This community, though rich in species, is in no sense indicative of old woodlands; it represents rather an assemblage of taxa which appears to have developed in well-lit situations where there is much enrichment in nitrates and phosphates, and is, in many localities today, a relic community which originated around pastures. With the introduction of modern chemical fertilizers into pastured parklands, it is tending to disappear, probably because nutrient concentrations become too high for it to persist on the bark. This community is most characteristic of elms and maples with bark of naturally high pH, but it can often be found on ash and even sometimes on old oaks where enrichment is adequate. The community occurs as a natural climax on the bark of species of oak in the relics of natural forest in Mediterranean Europe, where humidity is usually low, illumination high, and nutrient-rich dust impregnates the bark of trees freely, even in forests, in the dry summers.

In more sheltered situations, different epiphyte communities occur in pasture-woodlands, which are truly characteristic of old forest areas and, as a rule, do not occur in more recently planted woodlands. One can distinguish both succession in time and pattern in space in the epiphyte communities in such more humid sites. Well-lit twigs tend to bear communities of beard-lichens (*Usnea* spp.). The smooth bark of branches and of young trees tends to bear communities of crustose lichens that are normally fertile and are grouped in the phytosociological alliance of the *Graphidion.* Lichens such as *Graphis scripta* and *G. elegans* are characteristic of these habitats, often with cushion-forming mosses of the genera *Ulota* and *Orthotrichum.* With age, the bark of trees such as oak and ash becomes more fissured and spongy, and in long-undisturbed areas large foliose lichens of such genera as *Lobaria* eventually colonize the tree boles, usually in association with various mat-forming mosses such as *Isothecium myosuroides, Camptothecium sericeum* and *Hypnum cupressiforme.* This leads to the development of the very species-rich alliance known as the *Lobarion,* which seems to be the natural climax community, in one form or another, on mature trees in old pasture-woodlands, where there is adequate light. At this stage, bark seems to reach its highest pH (5.0-6.5), and also its maximum water-holding and base-exchange capacity. At this stage, too, the species of phorophyte seems to be less important; the community can develop, where the microclimate is suitable, on a wide range of tree species in lowland Britain, including oak, beech, ash, maples, elms and rowan. Conifers, however, never seem to carry this community in lowland pasture-woodlands and it is very rare on birch, sallow and alder.

The *Lobarion* alliance contains a very large number of lichen taxa, and as many as 30 or 40 may be found on a single old tree, though this number is exceptional. In more heavily shaded situations, lichens tend to be less prominent and bryophytes may form the main epiphyte cover.

Space. Most of the lichen communities so far considered are best developed on the south or south-west sides of tree boles, and also on the upper surface of horizontal branches, seemingly because of 2 factors. One is that such aspects receive the maximum solar energy, the afternoon sun tending to penetrate better than that in the morning when woods are often misty. The other is water supply. In Britain the prevailing rain-bearing winds are from the south-western quarter of the compass. If the bark surface is more or less horizontal, as it will tend to be on the upper side of a large bough, then water will tend to be retained there longer than on the sides or under-surface of such a limb. Often there are few or no epiphytes present in the last 2 positions.

The north and east sides of vertical trunks tend to carry shade-tolerant and less moisture-demanding taxa of lichens; in heavy shade with adequate, but not excessive, moisture (excessive moisture encourages bryophyte growth), such crustose lichens as *Enterographa crassa* and *Thelotrema lepadinum* may dominate in old pasture-woodlands on these sides of trees. Where, however, the tree leans towards the north or east, or there are underhangs rarely wetted by rain, specialized communities occur that do not seem to tolerate much direct water falling on or flowing over them. Here, on older trees especially, such species as *Lecanactis abietina, Schismatomma decolorans* and *Calicium viride* may occur. The pattern of rain-tracks on a trunk (which not only bring water but nutrients) is very important in controlling the pattern of lichen communities. The centres of rain-tracks are often colonized by bryophytes; their margins bear foliose lichen communities, often including *Lobaria* spp., while the drier areas further out bear crustose lichen communities. On very old trees, especially oak, changes occur in the bark. It becomes more brittle and less water-retentive and a post-climax community of crustose lichens may develop. One of the most characteristic of such communities is the *Lecanactidetum premneae* association found on very old oak bark, with *Lecanactis premnea, L. lyncea, Schismatomma virgineum,* etc.

When the bark finally falls from a tree through death, injury or disease, lignicolous lichen communities develop on the exposed wood, containing mainly species of *Calicium, Chaenotheca* and *Cladonia.* Once the wood becomes really soft and rotten, however, lichens disappear and the saprophytic fungi and Myxomycetes become plentiful.

It can be seen, therefore, that a single old pasture-woodland area may contain a great diversity of habitats, and hence many species of epiphytes. If only the lichens are considered here, an average of at least 100 corticolous and lignicolous taxa per square kilometre of woodland could be recorded for an ancient site with a moderate degree of diversification of ecological niches. Several sites reach a density of 140 taxa per square kilometre, while the richest sites may exceed 170: Boconnoc Park has 191 lichen taxa in a pasture-woodland area of about one square kilometre and Melbury Park has 213 taxa in a rather larger area. Such figures are far higher than for any comparable habitats of similar (or even greater) size known on the European continent, a fact which emphasizes the unique international importance of the British pasture-woodlands.

INVERTEBRATE HABITATS

The invertebrates particularly associated with old trees in pasture-woodlands exhibit a wide range of habitat requirements in their developmental stages. Although species such as the beetles *Cryptocephalus querceti* and *Rhynchaenus fagi* are associated, throughout the active stages of their life cycles, with the foliage of trees, many of the characteristic Coleoptera and Diptera are associated with the 'wood' of trees, whether this be the roots (eg *Prionus coriarus*) or thin twigs in the high canopy. Epiphytes themselves provide habitats for large numbers of insects, especially bugs (Hemiptera) and book-lice (Psocoptera). Morris (1974) described the habitats utilized by invertebrates in oak trees.

It is, however, the fauna of 'dead wood' that is the special feature of pasture-woodland invertebrates. Knowledge of this fauna is predominated by that of the Coleoptera and Diptera, but the total invertebrate fauna of dead wood is formidably large. Elton (1966) summarized the situation thus: 'Dying and dead wood provides one of the 2 or 3 greatest resources of animal species in a natural forest, and if fallen timber and slightly decayed trees are removed the whole system is impoverished of perhaps more than a fifth of its fauna'.

Four main structural types of dead wood were recognized by Stubbs (1972), after Elton (1966), all of which can occur in pasture-woodland areas. These are:

1. the living or partly dying standing trees;
2. boles after the tree crowns have collapsed;
3. logs of various sizes;
4. residual stumps.

Dead wood occurs within living trees. Larkin and Elbourn (1964) studied the fauna associated with dead wood in oak trees. Hunter (1977) gave a detailed account of the dead wood habitats available, for Coleoptera, within and surrounding a tree, although in this case the specific example was of a pine tree.

It is surprising that, although the invertebrate fauna of Britain must rank among the best documented in the world, at least in terms of occurrence and means of identification, comparatively little attention has been given to autecological studies. This lack is nowhere more apparent than in the case of the fauna of dead wood. Few species have been studied in any detail and knowledge of the precise larval and pupal habitat is, for many species of Coleoptera and especially of Diptera, only fragmentary. Such knowledge as does exist is mostly unpublished.

The succession of dead wood habitats and their various merits were reviewed, based on published and unpublished research, and upon anecdotal information, by Elton (1966), Stubbs (1972), Paviour-Smith (1973), Morris (1974) and Hunter (1977). Stubbs, in particular, was concerned with the conservation of dead wood fauna, drawing extensively on his own experience of the Diptera of Windsor Forest. A few authors, notably A A Allen, have contributed to our knowledge of the development and life histories of some of the uncommon Coleoptera associated with dead wood in pasture-woodland areas, but these contributions are widely scattered in the entomological literature of the last 100 years. The developmental ecology of some species has been studied by continental workers. Some of these studies (eg Palm 1959; Leseigneur 1972; Kelner-Pillault 1974) deal with species which occur in Britain.

Most records of Coleoptera and Diptera are based on finds of adult insects (imagines). Once the imago has emerged from the pupa, or puparium, it may disperse at some distance from the breeding site. Records of species taken in flight, on blossom, in what are probably day shelter sites (eg under bark) or at bait tell one little about the real habitat requirements of the insect. Many Coleoptera imagines appear to be nocturnal, or crepuscular, in habit, which makes it difficult to characterize their precise habitat requirements.

The very act of searching for the developmental stages of much dead wood fauna threatens the total fauna of the particular habitat being searched. The only way to search most dead wood habitats is to pull them apart, thereby destroying the structure. If one accepts the statement made by Stubbs (1972) that 'even in favourable localities it may only be one in a hundred large trees that provides the ideal conditions for any one given species ....', it is conceivable that certain species could be exterminated, or reduced to non-viable populations, by entomologists in some popular collecting localities.

Not all the invertebrates which are characteristic of old trees in pasture-woodlands feed directly on wood, its breakdown products or fungi. Some species are predatory. The pseudoscorpion *Dendrochernes cyrneus* and the spider *Lepthyphantes carri* are predatory within ancient trees throughout the active stages of their life cycles. Several Coleoptera are also predatory, either in the larval stage (eg *Elater ferrugineus*), or both as larvae and imagines (eg *Calosoma inquisitor*). Several Cleridae (Coleoptera) are predatory on the larvae of other Coleoptera. Some Diptera are parasitoids, such as *Xylotachina diluta* on the goat moth *Cossus cossus*. The beetle *Ctesias serra* feeds on insect remains in spider webs in hollow trees. The larvae of some Diptera are present as commensals in the burrows of wood-boring Coleoptera, eg *Odinia hendeli* with *Ischnomera caerulea,* and *Odinia meiferei* with *Scolytus scolytus.* Commensal Coleoptera also occur, especially with social Hymenoptera. The staphylinid *Velleius dilatatus* occurs with the hornet *Vespa crabro* which commonly nests in hollow trees. Several species occur in nests of the ant *Lasius brunneus*, itself a species largely restricted to pasture-woodlands, including *Euryusa optabilis, Batrisodes adnexus* and *Euconnus pragensis,* but such species often remain undetected until the tree is felled. Some solitary wasps (Hymenoptera) use the galleries created in dead wood by Coleoptera as nest sites.

Because of the inadequate information about the life histories of most of the invertebrates associated with old trees in pasture-woodlands, it is difficult to compare the habitat requirements of invertebrates and epiphytes. A fundamental difference is that the epiphyte flora is surface-dwelling, whereas the fauna is largely internal to dead wood. This contrast has parallels in the vegetational and faunal components of other ecosystems.

THE RELIC FLORA AND FAUNA OF PASTURE-WOODLANDS

EPIPHYTES

It has been suggested (Rose 1974, 1976; Rose & James 1974; James *et al.* 1977) that some epiphytes are both relics of the flora of the wildwood, and also indicators of continuity of elements of the structure and microclimate of the wildwood through a long period of time, without major breaks due to clear felling.

With occasional exceptions, these epiphytes (Table 4) are confined in lowland Britain to old pasture-woodland sites, or with those that have been managed in this way until recent times. Where they occur in sites no longer grazed, they correlate with lack of disturbance and persistence of large trees. A few such occurrences are in areas of coppice-with-standards, but in these cases the old standards are large and in some (eg East Dean Park, Sussex) the sites are former medieval deer parks.

The continuity of old trees at a site is, of course, often difficult to prove further back than the present generation of standing trees and existing stumps, but where they still occur it is often more likely to be the result of continuance of earlier traditional practice, rather than the reverse.

From the data given in Table 4, lists of taxa have been selected that correlate well with woodland sites known to be ancient in the New Forest and elsewhere, and also have a reasonably wide geographical distribution in England. These 'indicator species' have been used to calculate Indices of Ecological Continuity. The Revised Index of Ecological Continuity (RIEC) (Rose 1976) is based on the percentage occurrence of up to a maximum of 20 taxa out of a total of 30 taxa (see Table 5) in a site, and sites listed in Tables 8 and 9 have their RIEC indicated. It will be noted in both Tables that the RIEC values for certain sites, in fact, exceed 100%. This apparent anomaly arises because the RIEC was designed on a British national basis, rather than on a series of regional indices (Rose 1976). The RIEC includes some species found in certain parts of Britain and not in others; hence, all the species would rarely be expected to occur at one site, for climatic reasons. Several sites have more than 20 of the species listed in Table 5, so that values over 100% can occur at the richest sites where both temporal continuity of habitat and present-day habitat diversity are present. The most critical indicators of old forest continuity among the lichens appear to be the species of *Lobaria* and *Sticta*, while *Thelotrema lepadinum* is also useful

Table 4. Some lichen epiphytes faithful to old hardwood forest areas in lowland Britain, particularly to pasture-woodlands.

Agonimia octospora
Arthonia vinosa
A. stellaris
*Arthropyrenia ranunculospora**
Arthothelium ilicinum
Caloplaca herbidella
Bacidia biatorina
Bactrospora corticola
Biatorina atropurpurea
(= *Catillaria atropurpurea*)
Catillaria pulverea
C. sphaeroides
Catinaria grossa
C. laureri
Cetrelia olivetorum
Chaenotheca brunneola
Dimerella lutea
Enterographa crassa
Haematomma elatinum
Heterodermia obscurata
Lecanactis amylacea
L. lyncea
L. premnea
Lecanora quercicola
Lecidea sublivescens
Leptogium burgessii
L. cyanescens
L. teretiusculum
Lobaria amplissima
L. virens
L. pulmonaria
L. scrobiculata
Lopadium disciforme
Megalospora tuberculosa
(= *Bombyliospora pachycarpa*)
Nephroma laevigatum
N. parile
Ochrolechia inversa
Opegrapha sorediifera
O. ochrocheila
O. corticola
Pachyphiale cornea
Pannaria mediterranea
P. conoplea
P. rubiginosa
P. sampaiana
Parmelia arnoldii
P. crinita
P. minarum
P. horrescens
P. reddenda
Parmeliella triptophylla
P. jamesii
P. plumbea
P. testacea
Peltigera collina
P. horizontalis
Pertusaria velata
Phyllopsora rosei
Porina borreri
P. coralloidea
P. hibernica
P. leptalea
Pseudocyphellaria crocata
P. intricata
Pyrenula macrospora
P. chlorospila
Rinodina isidioides
Schismatomma niveum
*S. quercicolum**
S. virgineum
Stenocybe septata
Sticta limbata
S. sylvatica
Strangospora ochrophora
Thelopsis rubella
Thelotrema lepadinum

(Nomenclature follows Hawksworth *et al.* (1980), with 2 later changes. *These names are used by Coppins & James in press).

Table 5. Species of lichens used to calculate the Revised Index of Ecological Continuity (RIEC) (Rose 1976).

Arthonia vinosa	*Pachyphiale cornea*
Arthopyrenia ranunculospora *	*Pannaria conoplea*
Biatorina atropurpurea	*Parmelia crinita*
(= *Catillaria atropurpurea*)	*P. reddenda*
Catillaria sphaeroides	*Parmeliella triptophylla*
Dimerella lutea	*Peltigera collina*
Enterographa crassa	*P. horizontalis*
Haematomma elatinum	*Porina leptalea*
Lecanactis lyncea	*Pyrenula* (either *macrospora* or *chlorospila*, but do not count both)
L. premnea	*Rinodina isidioides*
Lobaria amplissima	*Schismatomma quercicolum* *
L. virens	*Stemocybe septata*
L. pulmonaria	*Sticta limbata*
L. scrobiculata	*S. sylvatica*
Nephroma laevigatum	*Thelopsis rubella*
	Thelotrema lepadinum

(Nomenclature follows Hawksworth *et al.* (1980), with one later change. *These names are used by Coppins & James in press).

as it is less sensitive and does not require continuity of very old trees. Lichens are rarely found as fossils, so that it is not possible to reconstruct directly the epiphytic flora assemblages of the wildwood.

INVERTEBRATES

Work done over the past 20 years, especially at the Department of Geology, Birmingham University, has provided information on elements of the insect fauna of the Flandrian forests of Britain, especially the Coleoptera associated with dead wood habitats. Hammond (1974) reviewed many of these data on post-glacial Coleoptera faunas in the context of overall changes in the coleopterous fauna of Britain. The possible effects of forest clearance by man on the coleopterous fauna of Britain were discussed by Osborne (1965), but much additional evidence from newly discovered fossil assemblages has been produced in recent years. A note of caution was sounded by Kenward (1976) on the use of fossil fauna death-assemblages for interpreting past ecological conditions, but the undeniable fact remains that many species of Coleoptera, particularly those known from their present occurrence to be associated with dead or dying wood, were preserved as fossil remains at various localities in Britain. It is improbable that many of these species constitute part of the background 'rain' of fauna described by Kenward (1976), which seems to be a counterpart of the background rain of pollen observed by many workers (see Godwin 1975).

It appears that several species of woodland Coleoptera present in the wildwood (eg *Cerambyx cerdo* and *Prostomis mandibularis*) have been lost (see, for example, Girling 1982), but Hammond (1974) considered that there are still approximately 650 'woodland species' of Coleoptera found in the British Isles. This list includes predatory ground beetles, such as *Calosoma inquisitor,* and the myrmecophilous species associated with the predominantly woodland-inhabiting ant *Formica rufa* (Donisthorpe 1913), as well as the more obviously tree-dependent species such as *Curculio venosus*, the larvae of which attack developing acorns, and the leaf-mining weevil *Rhynchaenus fagi.*

Some Coleoptera are particularly associated with overmature trees in wooded land in lowland Britain, as are some Diptera, spiders (Aranaea) and Pseudoscorpiones (Harding 1977a). The case for considering some of these as indicators of the continuity of woodland remains to be investigated. However, it is possible to conclude that, just as with epiphytes, there are several categories of invertebrates associated with pasture-woodlands in lowland Britain. Documentation of the occurrence of many species is fragmentary and the following observations refer to Coleoptera, especially the saproxylic species; however, presently available information suggests that the same principles will apply to Diptera and other groups. In Appendix 2, 196 species of mainly saproxylic Coleoptera are listed. These species have been grouped according to the extent to which they have been consistently recorded from areas of ancient woodland with continuity of dead wood habitats. The list has been modified from that given by Harding (1977a) in the light of a more comprehensive survey of published and unpublished records than that included in Harding (1978f).

Some species were doubtless constituents of the fauna of the wildwood, but like some epiphytes they are clearly able to adapt to modern conditions, and are able to colonize new woodlands or planted trees. Throughout most of lowland Britain, many species such as *Clytus arietus, Pyrochroa serraticornis* and *Anobium punctatum* occur commonly in a variety of tree species in all types of woodland, and even in hedgerow trees. The last species - woodworm - has been particularly successful, having become ubiquitous in dwellings, and *Clytus arietus* has been recorded breeding in seasoned timber in the open air, such as fence posts.

A further group (Appendix 2, Group 3) also contains a large number of species, and to some extent it contains species which are similar to the previous group, in that they can be found in all types of woodland and in isolated trees over parts of lowland Britain. However, in other parts of their range they are characteristic only of ancient woodland with overmature trees. For example, the lesser stag beetle *Dorcus parallelopipedus* is locally common in southern England and occurs particularly in ash in woods, hedges and other isolated trees. In northern England, it is much more restricted and occurs mainly in areas known to be ancient woodland, particularly pasture-woodlands. Conversely, *Hylecoetus dermestoides,* although generally much rarer than *Dorcus*, occurs widely in woodlands and plantations in the midlands and north of England and is extremely rare further south, being limited to a few old pasture-woodland localities (Shaw 1956; Allen 1966b). The occurrence, together, of numbers of species of this group is characteristic of ancient woodland areas.

Species which occur mainly in areas believed to be ancient woodland, but which also appear to have been recorded from areas that may not be ancient or for which locality data or site history are imprecise, comprise Group 2 in Appendix 2. This group includes species such as *Ischnodes sanguinicollis* and the myrmecophiles of the genera *Euryusa* and *Batrisodes* associated with the ant *Lasius brunneus.* Other species such as *Phloiotrya vaudoueri* and *Enicmus brevicornis* may, like those in Group 3, be restricted to ancient woodland areas over part of their geographic range. Another synanthropic species, death-watch beetle *Xestobium rufovillosum*, could possibly belong to this grouping as its occurrence away from old buildings is almost exclusively limited to ancient pasture-woodland areas (Buckland 1975, 1979).

The final grouping of species (Appendix 2, Group 1), those which probably qualify to be considered as indicators of old forest relic areas, is perhaps that which is most disputed by entomologists. In all cases, information is incomplete, but records of certain species are obviously limited to those from a few localities known to be ancient pasture-woodlands (eg New Forest, Windsor Forest, Sherwood Forest, Moccas Park). Some of these species are included in comparative lists for certain localities (Allen 1966a; Welch & Harding 1974). Since the collation of the original list of Coleoptera considered to be possible indicators of habitat continuity (Harding 1977a, 1978f), several authors have used the list to compare or evaluate the fauna of sites (Rose & Harding 1978; Hammond 1979; Welch & Cooter 1981; Atty 1983; Garland 1983).

The 68 species listed in Group 1 of Appendix 2 probably constitute part of the British assemblage of old forest animals (Urwaldtiere) discussed by Palm (1959), Buckland and Kenward (1973), Harding (1978f) and Hammond (1979).

The *British Red Data Book* for insects (Shirt 1986) listed 550 species of Coleoptera whose survival in Britain is considered to be threatened. The *Red Data Book* covered the Coleoptera of all habitats, but of the 'Endangered' and 'Vulnerable' species the most important was woodland (40% of species in these categories). Many of the species listed in Appendix 2 are included in the *Red Data Book,* with a high proportion being considered 'Endangered' or 'Vulnerable'. Seventy-four species of saproxylic Coleoptera, drawn from Groups 1 and 2 of Appendix 2, which are also included in the *Red Data Book,* are listed in Appendix 3. The occurrence of these species at 11 sites is compared to demonstrate the importance of pasture-woodland sites, such as Windsor Forest, the New Forest and Moccas Park, for the conservation of these specialized and threatened insects.

PASTURE-WOODLANDS IN RELATIONSHIP TO OTHER WOODLAND TYPES

Writing in the context of wildlife conservation priorities, Peterken (1977a,b) summarized 5 types of woodland which possess special scientific interest. These types were modified from Peterken (1974) to obviate the need to distinguish between primary and ancient secondary woodland. One of Peterken's types is pasture-woodlands (described as 'wood-pasture'). Three of the other types can be considered in relation to pasture-woodlands in lowland Britain in the context of the conservation of epiphytes and saproxylic invertebrates. These are (i) ancient coppice-woodlands, including the structural variants resulting from management, (ii) ancient woods in topographically inaccessible sites, (iii) woods formed by at least 150 years of largely natural succession and structural development.

Most coppice- and coppice-with-standards woodlands, even in areas of low air pollution, have extremely limited epiphyte floras, with few taxa and low RIEC values, even where they are known to be primary woodlands. The coppice cycle appears inimical to the maintenance of a rich epiphyte flora. Not only are standard trees often removed before they reach a large size, but the mature coppiced shrubs shade very heavily the boles of the remaining standards. After the coppice is cut, there is a phase of good illumination of the boles of the standard trees, but this is sudden, causing drying out, and is also brief. Such conditions clearly were not found in primeval forests and seem intolerable to many epiphytes. Where large old standards and occasionally pollards have been permitted to remain along internal boundaries and rides, however, a richer epiphyte flora may persist, including some of the apparently more tolerant crustose old-forest lichens such as *Pachyphiale cornea* and *Arthonia vinosa,* but rarely the less tolerant foliose taxa such as *Lobaria* spp.

Few coppice-, or coppice-derived, woods have been surveyed in any detail for invertebrates. Those that have, and for which lists are available, seem to lack, or to have few of, the highly adapted old-forest species associated with the dead wood biocoenosis. One of the most intensively surveyed and well-documented coppice-derived woods, for Coleoptera, is Monks Wood, Cambridgeshire (Welch 1973, 1968-80), from which 5 of the 74 species listed in Appendix 3 are recorded. This number of species compares favourably with the records (derived mainly from published sources and therefore probably incomplete) for certain pasture-woodlands listed in Appendix 3. However, in terms of completeness of survey, only a few pasture-woodlands compare with Monks Wood. Several well-surveyed woods have none, or only one or two, of the species in Groups 1 and 2 of Appendix 2 recorded for them, eg Bedford Purlieus (Welch 1975), because of an almost total lack of old or even of mature broadleaved trees.

Woods in topographically inaccessible sites, such as various inland cliffs and sea cliffs, are mainly located in the west of Britain. Because of this geographical location they are often rich in epiphytes, including species indicative of ecological continuity. Several woodlands of the type are included in Rose (1976, Table I). Few entomological records from woods of this type exist, but the available information suggests that the saproxylic fauna is sparse. In the Upper Clyde valley, however, certain gorge woodlands are known to contain a few species of Coleoptera believed to be indicative of old-forest conditions (Crowson 1962, 1964).

Enclosed woods formed by a long period of natural structural development have been examined for epiphytes, and records of invertebrates also exist for some sites. Indications are that, although inferior to pasture-woodlands, some areas, notably those with a history of woodland continuity, do contain elements of the old-forest epiphyte flora and saproxylic fauna. Enclosed woodlands known to have been plantations, even where they are now mature high forest, have much poorer densities of epiphyte taxa than the old, or former, pasture-woodland sites considered earlier. They also have very low values of the RIEC. This applies, for example, to the planted high-forest areas of oak and beech in Lakeland, the Forest of Dean, and the Weald. Some of the New Forest plantations are rather richer both in total epiphytes and in their RIEC values: such sites are always adjacent to ancient pasture-woodlands in the forest, whence presumably some colonization has been possible.

(Photograph F Rose)

PLATE 1
Brampton Bryan Park, Hereford and Worcester. Ancient oak with a rich epiflora of lichens, bryophytes and polypody ferns. The hole in the fork of the tree almost certainly provides access to an area of heart-rot providing favourable habitat for several beetle species.

PLATE 2
Moccas Park National Nature Reserve, Hereford and Worcester. Heart-rot in a living oak exposed when half the tree blew down in a gale. This type of reddish heart-rot within a tree provides habitat for several very rare beetles at Moccas, but once exposed like this soon desiccates and becomes almost sterile.

(Photograph P T Harding)

(Photograph F Rose)

PLATE 3
Lobaria pulmonaria is one of the most distinctive lichens of ancient woodlands in pollution-free areas. It is more tolerant of dry conditions and high light levels than other *Lobaria* spp and is often found on old trees in ancient pasture-woodlands.

PLATE 4
Lobaria amplissima is similar in most of its requirements to *L. pulmonaria,* but is less widely distributed. A typical member of *Lobarion pulmonariae* community.

(Photograph F Rose)

PLATE 5

Thelotrema lepadinum. In lowland Britain this species is a valuable indicator of ancient and undisturbed woodlands, especially pasture-woodlands. It favours the smooth-barked boles and branches of beech, oak, hazel and holly.

(Photograph F Rose)

(Photograph F Rose)

PLATE 6

Nephroma laevigatum is a characteristic member of the *Lobarion pulmonariae* community of epiphytic lichens, found mainly on oak, ash, hazel, rowan and wych elm.

(Photograph J Mason)

PLATE 7
Dorcus parallelopipedus. The lesser stag beetle is common and widespread in southern England, but in the north it is mainly restricted to ancient woodland, especially pasture-woodlands.

PLATE 8
Ampedus cinnabarinus. One of the uncommon red click beetles which has been recorded in recent years mainly from pasture-woodland areas such as the New Forest, Forest of Dean and various parks and woods in Sussex.

(Photograph R C Welch)

CONSERVATION

CONTINUITY OF THE HABITAT

Several factors affect the likelihood of continuity at individual sites.

Phorophyte/host tree species

A succession of age classes of appropriate species must always be present. Information is not available on how long an individual tree is capable of remaining in an 'over-mature' state, or how much overlap of generations is necessary. Many pasture-woodlands, now with abundant overmature and mature trees, lack a young generation of trees to provide continuity in the future. At a few sites no young generation exists and the gap between generations, even if young trees were to be planted or enclosures made now, may already be too great for future continuity to be maintained, although selective pollarding of some trees could extend the age range of the tree stock.

Structural management

Pasture-woodlands are an unfashionable form of land management in lowland Britain. Consequently, many sites are being subjected to changes of management which adversely affect their conservation value. The commonest change is the opening up of the woodland to increase the area for pasture or, usually more drastic, for arable farming (eg Bagot's Park in Staffordshire). Temple Newsam Park in Yorkshire, a recently discovered, important entomological site, has almost completely disappeared as a result of open-cast coal-mining. Recreational use of pasture-woodlands, particularly parks, is increasing, with sites being used as golf courses (eg Lullingstone Park in Kent (see Pitt 1984)), safari parks (eg Longleat Park) and public leisure areas, especially Country Parks (eg Bradgate Park in Leicestershire), usually resulting in the removal of the older trees. Conversely, management has changed at some sites to convert pasture-woodland to plantation woodland (Savernake Forest, Windsor Forest), with the result that overmature trees die prematurely, because of shading, and the discontinuous canopy typical of pasture-woodland is eventually replaced by closed canopy of high forest, which will be felled at maturity (in forestry terms), before many dead-wood habitats have developed.

Pollarding of trees, which had been so effective in prolonging the lives of trees and thereby providing suitable habitats, is an obsolete form of management which has not been practised on any scale in pasture-woodlands for over a century.

Dead wood

As described earlier, the natural death and decay of wood provide habitats for many invertebrates. Almost all pasture-woodlands suffer from the removal of dead wood, usually for local use as fuel. Frequently so much dead (and dying) wood is removed that dead-wood habitats almost cease to exist, particularly in areas where public access is allowed or encouraged. For example, in some Country Parks, 'tree surgery' is done on any dead branch or hollow tree (eg Studley Royal Park in North Yorkshire, when managed by the local authority) and 'unsafe' or 'unsightly' trees are removed. The importance and management of dead wood were reviewed by Stubbs (1972).

Atmospheric pollution

In the case of epiphytes, atmospheric pollution poses an additional threat to populations. Midland and East Anglian sites are much at risk from increasing rural air pollution due to the present policies of building tall chimneys at electric power stations. Such chimneys have reduced urban air pollution, but have tended to increase air pollution widely over rural areas. Apart from the coastal belt from north Norfolk to north-east Suffolk, the outlook for the epiphyte communities in East Anglia and in the Thames valley is rather poor and a decline in biomass and in richness has been observed in several sites (eg Ickworth Park and Windsor Forest) over the last 15 years.

Other factors increase the vulnerability of sites, particularly in midland and eastern England, north of the Weald. These factors include the relatively dry climate, which accentuates the effect of 'opening up' woodlands and of drainage within woodlands; lowering of the water table because of water extraction, for domestic supplies or as a result of mining; and the increasing use of agricultural chemicals, both fertilizers and pesticides, which become increasingly wind-borne into at least the marginal zones of pasture-woodlands bordered by agricultural land. Even neglected pasture-woodlands may be adversely affected - by excessive natural regeneration leading to the shading of old trees by young ones.

In the 'Ancient and Ornamental Woodlands' of the New Forest, natural regeneration is in general adequate, and the continuance of the habitats seems assured for at least the near future, as long as air pollution does not increase, no excessive woodland drainage takes place, and recreational pressures are restricted. Sites in south-west England and in Wales, in regions of more abrupt relief, higher rainfall and general air humidity, especially those in valleys such as Boconnoc Park, Horner Combe, and Arlington Park in Devon, also have good prospects for continuity, provided they are not disturbed unduly by felling of more mature trees, excessive opening up of the canopy, underplanting, or the uncontrolled natural regeneration of invasive species such as sycamore.

VIABILITY OF POPULATIONS

In the belt from the Weald to Somerset and then northward through Wiltshire, Herefordshire and east Wales, and from the north Pennines and Lakeland to south-west Scotland, the populations of most uncommon epiphytes, though often small and in some cases sterile, seem to be viable in many sites at present. They are relic in the sense that they seem unable to colonize new habitats, but apparently can maintain themselves, in many cases largely by vegetative reproduction, in their existing habitats. In many sites from Exmoor and west Dorset to Cornwall and locally in west Wales, the luxuriance and abundance of the larger foliose species are greater, and more (though not all) of the species are fertile. Here the situation is still better, though some species are nevertheless very rare. For example, *Lobaria amplissima* is at present known on only 41 trees in the whole of England, on 25 trees in Wales, and on perhaps only 6 in Galloway, south-west Scotland - a total of 72 trees in Britain south of the Scottish highlands. *Lobaria scrobiculata* was recorded as fertile on one old tree in St Leonards Forest, Sussex, by W Borrer (Turner & Dillwyn 1805) at the beginning of the 19th century. It is now known to be fertile in Great Britain only in the old woodlands of the west highlands of Scotland, from Argyll to Ross, and is now extinct in England, except in Cornwall, Devon and the Exmoor area of western Somerset, although it is surviving in western parts of Wales. *Evernia prunastri* was known in the fertile state early last century, as near to London as Virginia Water, Surrey. It is still widespread and common throughout most of England away from large cities and industrial zones, but almost always in the sterile state. The only sites in south-east England where this species has been recorded with ascocarps in the last 10 years are near Hythe in Kent and in east Dorset.

It is clear, however, that a decline in fertility of many lichen and bryophyte taxa does not preclude their continued survival at a site by vegetative reproduction, but in the case of some *Lobaria* species, it may well be responsible for their apparent failure to colonize more recently created potential habitats (such as mature broadleaved plantations) at some distance from existing sites. Many species of invertebrates associated with dead wood seem to have limited powers of dispersal. A few of the Coleoptera are flightless (eg *Batrisodes* spp.) and others seem to be active only in favourable climatic conditions (high temperature and humidity), often at dusk. In the more or less continuous primeval forest, a limited ability to disperse would not have seriously disadvantaged a species, but with the isolation of most pasture-woodland areas today, it is a major disadvantage. The isolation of site from site in existing pasture-woodlands, especially parkland, poses a problem in itself. However, the increasing isolation of trees within sites which have been opened up for agriculture or recreation compounds the effects of limited dispersal abilities.

An example of this isolation is demonstrated by the beetle *Hypebaeus flavipes,* which is known to breed in Britain only at Moccas Park. It has been recorded from only about 6 ancient oak pollards (J Cooter, pers. comm.), but not from any of the overmature maiden oaks nearby.

It is evident that some species have remained at some well-recorded sites, presumably at very low densities, for numbers of years, without being detected by collectors. Several examples among the Coleoptera at Windsor Forest were described by Allen (1966a) and more recently by J A Owen (1984 and pers. comm.). The most notable examples are probably *Lacon querceus,* recorded before 1830 and not again until 1936, subsequently to be recorded in several years; and *Ampedus nigerrimus* which occurred sporadically in less than 20 individual years between 1841 and the 1960s. The problem of over-collecting rare species, or of inadvertent habitat destruction while collecting at particularly favoured sites, such as the New Forest, has been mentioned earlier, but, without some knowledge of the life histories and population dynamics of the species concerned, it is almost impossible to predict the effects of these activities.

STATUTORY CONSERVATION

In his chapter on 'The selection of nature reserves' Sheail (1976) drew upon a wealth of unpublished and otherwise unexploited information about the history of the selection of areas considered to be important for wildlife conservation, up to the 1940s. Little of Sheail's account dealt directly with pasture-woodlands, but his account of the selection of potential nature reserves, made in 1915, 1929 and 1947, is of relevance.

In 1915, the Society for the Promotion of Nature Reserves (SPNR) selected 251 areas (of all types of habitat) as being 'worthy of protection' for wildlife. This list was compiled mainly by N C Rothschild and G C Druce in a laudable effort to forestall the clearance of areas of conservation value, to make way for war-time agriculture. Sheail (1976, pp 129-131) named only a few of these areas, but it is clear that, although the SPNR had 'an eye to Europe as a whole' by selecting sites which included habitat types with no exact counterpart in continental Europe, pasture-woodlands were neglected. The New Forest appears to be the only pasture-woodland area included, and this was considered to be 'in no immediate danger of being destroyed' because 'various Acts of Parliament regulating its management preserved to some extent the natural features of the locality'.

When presenting its evidence to the Addison Committee in 1929, the British Correlating Committee stated that 'nature reserves are most required' in 66 areas, some very extensive like the New Forest, and others very small, eg Martlesham Common in Suffolk. Apart from the New Forest, the importance of which was obviously realized (although whether for wildlife or recreation is not clear), several large forest areas are listed (Sheail 1976, pp 130-134). These included, in most cases, some areas of relic pasture-woodland with old trees. The Forests of Dean, Wyre, Savernake, Pamber, Ashdown and Charnwood, and Cannock Chase were included. Surprisingly, Windsor and Sherwood Forest, both well known at that time for Coleoptera, were not included, despite the fact that representatives of the Royal Entomological Society contributed to the list. At the same time, 1929, concern was being expressed by naturalists, especially entomologists, in southern England about the current and future management of the New Forest (Haines 1929).

The Addison Committee report was not implemented by the Government and the booklet published by R B Burrows in 1938, advocating the designation of 4 nature reserves which would include the major habitat types in Britain, aroused comparatively little interest. It is interesting to note that Burrows suggested that a woodland reserve might include the remnants of old woodland in the New Forest, Sherwood or Charnwood Forest, or the Forest of Dean, and that he selected these areas, typical of ancient forest, rather than the coppice-woodlands of greater interest for phanerogamic plants.

The Nature Reserves Investigation Committee set up in 1942 took evidence on proposed reserves from the Royal Society for the Protection of Birds, British Ecological Society and Royal Entomological Society, and from regional sub-committees. The Huxley Committee drew heavily on this information in compiling its list of proposed nature reserves and scientific areas in 1947. Sheail (1976, pp 150-151) listed and mapped the location of these areas. The New Forest was included as a scientific area, within which the Matley and Denny area was proposed as a nature reserve. Other pasture-woodlands included in the list of proposed nature reserves were Windsor Forest, Burnham Beeches and Epping Forest. It is ironic to note that Sheail (1976, pp 149-152) recorded that Staverton Park and Thicks in Suffolk was rejected by the Huxley Committee on the evidence of E J Salisbury, who asserted that the trees there were a remnant of an old plantation rather than an ancient woodland. Evidence for the historical continuity of woodland on the site was presented by Peterken (1969) and it is now regarded as an important pasture-woodland site, rich in epiphytes (Rose, in Peterken 1969) and in invertebrates associated with the old trees, especially Coleoptera (Welch & Harding 1974). The lists provided by the Huxley Committee in 1947 for England and Wales (and by the Ritchie Committee in 1949 for Scotland) were used as a basis for reserve acquisition by the Nature Conservancy. This body was set up by royal charter in 1949 and given legal powers under the National Parks and Access to the Countryside Act of 1949.

During the existence of the Nature Conservancy and its successor organization, the Nature Conservancy Council, formed in 1973, the protection for wildlife conservation of pasture-woodlands has been limited mainly to the areas which are scheduled as Sites of Special Scientific Interest (SSSI). Parts of the New Forest (Matley and Denny, Mark Ash and Bramshaw) and of Windsor Forest (High Standing Hill) were, for a time, protected as Forest Nature Reserves. This category of reserve no longer applies to these areas, because semi-formal agreements about the management of the whole of the New Forest and Windsor Forest have been made by the Nature Conservancy Council with the Forestry Commission and the Crown Estates Commissioners, respectively. Three areas, considered to be former pasture woodlands, are designated as National Nature Reserves, and a further area is designated as a Local Nature Reserve (Table 6). Moccas Park was declared a National Nature Reserve, under a nature reserve agreement, in 1981. In addition, The Mens and Ebernoe Common in Sussex are now owned by the Sussex Trust for Nature Conservation and managed as nature reserves (Tittensor & Tittensor 1977).

It is probably true to say that few ecologists or conservationists (other than some entomologists) accepted the importance of pasture-woodlands until comparatively recently. It has been mainly as a result of work done by a small group of botanists interested in epiphytes that the most up-to-date selection of potential nature reserve areas (Ratcliffe 1977) included 37 areas of pasture-woodland, some of which were totally unknown or neglected until recently. Of these 37 areas, 21, all in England, were listed as being 'mixed deciduous woodland: ancient parks and overmature woodland', the remaining 16 areas being listed as other woodland types. Perhaps more than any other woodland type, it is difficult to give a precise definition of the biotope and of its associated flora and fauna. However, the selection in Ratcliffe (1977) is based mainly on botanical information available before 1970, and subsequent work on the epiphytes and invertebrates of pasture-woodlands has caused some reappraisal to be made.

A list 68 of the most important pasture-woodlands, as known to date, is included in Table 6. The criteria for selecting the sites listed in Table 6 are: (i) known richness and diversity of epiphyte flora and/or saproxylic invertebrates; (ii) structural suitability for saproxylic invertebrates (and to a lesser extent for epiphytes) supported by historical evidence of habitat continuity; (iii) good prospects for the maintenance of structural continuity, although future management trends cannot be predicted at most sites. The list in Table 6 includes 34 of the 37 areas listed by Ratcliffe (1977).

Table 6. British lowland pasture-woodlands of importance for wildlife conservation

Site Name	County	National Grid Reference	Total number of epiphytic lichen taxa	RIEC value	Total number of threatened saproxylic Coleoptera recorded (see Appendix 3)	Legislative conservation status (a)	NCR grade (Ratcliffe 1977)(a)	Conservation priority rating (b)	Notes
ENGLAND									
Windsor Forest/Park	Berkshire	41/8.6. and 41/9.7.	75	20	54	SSSI	1	1	6
Burnham Beeches	Buckinghamshire	41/95.85.	50	25	-	SSSI	2	3	2
Boconnoc Park & Woods	Cornwall	20/14.60.	191	145	-	SSSI	1*	1	2
Eden Gorge	Cumbria	35/52.43.	70	20	-	SSSI	-	3	2
Great Wood, Borrowdale	Cumbria	35/27.21.	122	120	-	SSSI	1	2	2.7
Lowther Park	Cumbria	35/52.24.	80	20	-	-	2	3	2
Low Stile Wood, Borrowdale	Cumbria	35/23.12.	174	125	-	SSSI	1	2	2.7
Rydal Parks	Cumbria	35/36.05.	122	65	-	-	-	3	2
Yew Scar, Gowbarrow	Cumbria	35/41.20.	116	100	-	SSSI	2	2	2.7
Chatsworth Park	Derbyshire	43/25.70.	16	0	1	-	-	3	4.6
Arlington Park	Devon	21/62.40.	143	115	-	-	-	2	2.7
Brownsham/Clovelly Park Woods	Devon	21/29.25. to 21/30.25.	133	125	-	SSSI	2	2	2
Holne Chase	Devon	20/71.72.	139	115	-	SSSI	1*	2	2
Lustleigh Cleave & Bovey Valley	Devon	20/75.81.	154	115	-	SSSI/NNR	1	2	2
Shute Deer Park	Devon	30/24.97.	93	65	2	-	-	3	4
Ugbrooke Park	Devon	20/87.77.	77	45	-	-	-	3	2
Walkham Valley	Devon	20/54.73.	152	105	-	SSSI	-	2	2
Cranborne Chase	Dorset/Wiltshire	31/9.1.	166	105	-	SSSI	2	2	2
Holt Forest	Dorset	41/03.05.	40	35	-	SSSI	-	3	2

Lulworth Castle Park	Dorset	30/85.82.	123	70	-	SSSI	-	3	2
Melbury Park	Dorset	31/56.05.	213	120	-	SSSI	1	1	2
Great & Shipley Woods	Durham	45/00.21.	108	55	-	SSSI	1	3	2
Epping Forest	Essex	51/4.9.	38	10	13	SSSI	2	2	6
Harewood Forest	Hampshire	41/40.43.	48	10	3	-	-	3	4
New Forest	Hampshire	See Table 7			34	SSSI	1*	1	4
Brampton Bryan Park	Hereford/Worcester	32/35.71.	130	55	1	SSSI	1	2	4
Moccas Park	Hereford/Worcester	32/34.42.	109	25	19	NNR	1	2	
Cobham Park & Woods	Kent	51/69.68.	39	0	3	SSSI	-	3	4.5.6
Knole Park	Kent	51/54.54.	90	35	4	SSSI	-	3	4
Lullingstone Park	Kent	51/51.64.	55	35	1	-	-	3	4
Mersham Park	Kent	61/06.40.	104	40	-	SSSI	-	3	2
Bradgate Park	Leicestershire	43/53.11.	-	-	-	SSSI	-	3	4.6
Donington Park	Leicestershire	43/41.26.	-	-	-	-	-	3	4.6
Grimsthorpe Park	Lincolnshire	53/02.20.	29	0	2	SSSI	-	3	6
Richmond Park	London	51/1.7. & 51/2.7.	8	0	13	-	-	3	6
Dunham Park	Manchester	33/74.87.	-	-	7	SSSI	-	3	6.7
Whitfield Park	Northumberland	35/77.55.	108	40	-	SSSI	2	3	
Duncombe Park	North Yorkshire	44/58.83.	-	-	1	-	-	3	1.4
Sherwood Forest	Nottinghamshire	43/5.5. & 43/6.6.	8	0	19	SSSI	2	2	6
Blenheim Park	Oxfordshire	42/43.16.	96	15	4	SSSI	-	3	4
Wychwood	Oxfordshire	42/33.16.	128	40	-	NNR	1	3	4
Horner Combe complex	Somerset	21/89.44.	166	125	-	SSSI	1	1	2.7
Mells Park	Somerset	31/71.48.	120	60	-	-	-	3	2
Barle Valley Woods	Somerset	21/8.3.	157	130	-	-	-	2	2
Brockton Coppice	Staffordshire	33/98.19.	-	-	1	SSSI/LNR	2	3	1.4.6
Sotterley Park	Suffolk	62/46.85.	92	20	-	SSSI	1	3	2
Staverton Park	Suffolk	62/35.50.	65	35	5	SSSI	1	3	
Packington Park	West Midlands	42/22.84.	-	-	-	-	-	3	1.2.6
Ashburnham Park	East Sussex	51/69.14.	160	65	-	SSSI	2	3	2
Eridge Park	East Sussex	51/56.35.	178	85	-	SSSI	1*	2	2
Arundel Park	West Sussex	50/01.08.	86	30	6	SSSI	-	3	4
Ebernoe Common	West Sussex	41/97.26.	111	50	-	SSSI	1	3	2
Parham Park	West Sussex	51/06.14.	165	60	2	SSSI	2	2	4
The Mens	West Sussex	51/02.32.	77	40	-	SSSI	1	3	4
Longleat Park & Woods	Wiltshire/Somerset	31/81.43.	161	90	-	SSSI	2	2	2
Savernake Forest	Wiltshire	41/23.66.	112	60	1	SSSI	2	3	4

Site Name	County	National Grid Reference	Total number of epiphytic lichen taxa	RIEC value	Total number of threatened saproxylic Coleoptera recorded (see Appendix 3)	Legislative conservation status (a)	NCR grade (Ratcliffe 1977)(a)	Conservation priority rating (b)	Notes
SCOTLAND									
Glenlee/Garroch Woods	Dumfries & Galloway	25/5.6.,25/6.7.,25/6.8.	131	105	-	SSSI	-	2	4
Knockman & Garlies Woods	Dumfries & Galloway	25/41.68.	126	65	-	-	-	3	2
Lochwood	Dumfries & Galloway	35/08.97.	115	75	-	SSSI	-	3	4
Dalkeith Old Oakwood	Lothians	36/33.68.	67	0	-	SSSI	-	3	4
Cadzow & Avon Gorge	Strathclyde	26/72.52.	-	-	1	SSSI	2	3	1.4
Kilkerran	Strathclyde	26/31.03.	122	95	-	-	-	3	2
WALES									
Coedmore	Dyfed	22/19.43.	120	70	-	SSSI	-	2	2
Dynevor Park & Woods	Dyfed	22/61.22.	154	85	-	SSSI	-	3	2
Coed Crafnant	Gwynedd	23/61.29.	167	125	-	SSSI	1*	1	
Coed Ganllwyd	Gwynedd	23/72.24.	115	115	-	NNR	1*	1	
Gregynog Park	Powys	32/08.97.	148	65	-	SSSI	-	3	2
Powis Castle Park	Powys	33/21.06.	-	-	-	-	-	3	1.2

Notes on Table 6

Legislative conservation status

NNR - National Nature Reserve
(declared under section 19 of the National Parks and Access to the Countryside Act 1949 or section 35 of the Wildlife & Countryside Act 1981)

LNR - Local Nature Reserve
(declared under section 21 of the National Parks and Access to the Countryside Act 1949)

SSSI - Site of Special Scientific Interest
(notified under Section 23 of the National Parks and Access to the Countryside Act 1949 or section 28 of the Wildlife & Countryside Act 1981)

Indication that a site has some legislative conservation status or is listed by Ratcliffe (1977) does not necessarily mean that the whole of the site is so designated. In many cases only a part may be an SSSI, etc, or be listed. For the meaning of the grades in the *Nature Conservation Review* (1*, 1, 2) see Ratcliffe (1977).

Conservation priority rating - based on present knowledge

1 - Internationally important site
2 - Nationally important site
3 - Regionally important site

Some seemingly anomalous sites are listed in Table 6. A few historically ancient pasture-woodlands, such as Cranborne Chase, are now managed mainly as coppice or high forest, and some parks originated from agricultural land with hedges and woods, in heavily wooded areas (eg Sotterley Park). These areas now possess, probably by retention rather than by acquisition, a flora and/or fauna characteristically associated with ancient pasture-woodland.

Notes

1 - Lists of epiphytic lichens not available
2 - Lists of Coleoptera not available
3 - Only incomplete lists of epiphytic lichens available
4 - Only incomplete lists or isolated records of Coleoptera available
5 - Records of Coleoptera mainly from the 19th century
6 - Epiphytic lichen lists are available, but the site is subject to high levels of atmospheric pollution
7 - Owned or managed by the National Trust

The New Forest - a special case

The New Forest, Hampshire, is clearly a special case. None of its woodlands are primeval in the sense that they have never been disturbed by at least some degree of selective felling, management for deer production, or excessive grazing at times by ponies or cattle. It appears to contain some areas that have been less altered in their structure, and in their continuity of mature or old trees, than any other area of comparable size in western lowland Europe. After considerable survey work in Europe (FR), it appears that the New Forest contains the largest area of pasture-woodland remaining in north-west Europe. Flower (1977, 1980) made an historical study of the New Forest in relation to its present structure.

The New Forest is of unique importance internationally for epiphytic lichens, with 300 corticolous and lignicolous taxa still present in its unenclosed woodlands which total some 3600 ha (9000 acres) today. It has an unique concentration of 'old-forest indicator' epiphytes, especially of lichens, but also of bryophytes (Rose & James 1974). Table 7 gives details of the location and lichen epiphyte content of 13 of the most important areas of ancient pasture-woodland in the New Forest.

Entomologically, the New Forest is also considered to be of unique importance for a number of taxonomic groups and for a number of habitat types. The saproxylic fauna is poorly documented, although a manuscript list of the Coleoptera does exist (Gardner *et al.*). Using this list, the New Forest is slightly inferior to Windsor Forest in terms of the species listed in Appendix 3. However, many parts of the New Forest which contain overmature trees have apparently never been surveyed, either because they were unknown to entomologists, or are in areas remote from public roads.

Table 7. Sites of major importance for epiphytic lichens in the New Forest. (All sites are old pasture-woodland)

	Name	National Grid reference	Number of lichen taxa per km^2	RIEC
A	**Beech-dominated woods**			
	*Stricknage Wood	41/26-12-	133	95
	Wood Crates	41/27-08-	159	95
	*SW part of Busketts Wood	41/30-10-	171	105
	E part of Mark Ash Wood	41/24-07-	167	115
B	**Beech-oak mixtures**			
	Bramshaw Wood	41/25-16-	154	115
	Great Wood, Bramshaw	41/25-15-	148	105
C	**Oak-dominated woods**			
	Stubbs-Frame Wood	41/35-03-	168	110
	*Sunny Bushes	41/26-14-	88	90
	*South Brinken Wood	41/28-05-	88	90
	*South Ocknell Wood	41/24-11-	144	120
	Pinnick Wood	41/19-07-	108	80
	Red Shoot Wood	41/18-08-	168	90
	Hollands Wood	41/30-04-	140	100

*Areas of considerably less than 1 km^2.

CONCLUSIONS

The conservation of woodlands in Britain, for wildlife, has been directed mainly towards those sites with rich and diverse arrays of flowering plants, or towards presumed examples of vegetational types. This is an acceptable principle, but does tend to neglect the fact that few, if any, woodlands in lowland Britain have escaped some modification by man since the Flandrian forest cover became fully developed.

There is a developing awareness that the present-day flora and fauna of woodlands in Britain, as with other habitat types, are there as a result of man's influence on natural processes, and that the selection of areas for wildlife conservation should be made with this fact in mind. The question of the selection of nature reserves in this context is discussed by Tittensor (1981). The influence of present and past management practices on the criteria for selecting areas of woodland for wildlife conservation was discussed by Peterken (1974, 1977a, b). One of the 5 management types identified by Peterken (1977a) was 'relics of medieval wood-pasture' (pasture-woodland) management.

It is almost only in pasture-woodlands that the overmature tree/dead-wood spectrum of the primeval forest ecosystem survives in present-day lowland Britain. Ancient pasture-woodlands which retain something of their former structure, with overmature deciduous trees, are of high conservation value because they contain elements of the flora and fauna believed to be relics of those found in the primeval Flandrian forest and which are not, or only rarely, found elsewhere.

This relic flora and fauna is composed of cryptogams, particularly epiphytic lichens and bryophytes, and of invertebrates associated with the timber of overmature trees and with the dead-wood biocoenosis. The habitat requirements of this flora and fauna include features which, by inference from, and by knowledge of undisturbed primeval forest elsewhere in temperate regions (see Elton 1966), would have been present in the primeval forest cover of Britain. These assemblages of species have not been found consistently in other types of woodland in lowland Britain, or in other habitat types.

The flora and fauna associated with trees in pasture-woodlands include species assemblages which are probably more typical of primeval forest than the flora and fauna of the trees in other types of woodland management. This is because in pasture-woodland the internal environmental regimes, in terms of humidity and light, are relatively stable, and are related on a long timescale to the natural life-span of the canopy-forming trees. Also, there is far more stability and continuity of habitat in the trees themselves. In other woodland types, such as coppice-woodland, the humidity, light and substrate regimes fluctuate violently during the relatively short timescale of the management cycles.

Ancient pasture-woodlands, such as those listed in Tables 6 and 7, merit preservation in the interests of wildlife conservation. Few examples of pasture-woodland are protected so that they are managed in the interests of nature conservation. Pasture-woodlands were poorly represented, as compared with other woodland types, in the *Nature conservation review* (Ratcliffe 1977), and the selection was, necessarily, based on incomplete information. It is now certain from recent intensive fieldwork by one of us (FR), that the pasture-woodland management type is better represented in lowland Britain than in most of north-western Europe: the responsibility to conserve such areas is therefore not only regional or national, but also international.

Legislative conservation is only part of the process of conserving these areas for wildlife. The management of most areas must be considered. However, hastily considered changes of management should be resisted until research has demonstrated the most appropriate course of action. Far too little is known about the age structure of trees in pasture-woodland, the rates of turnover or what would be adequate regeneration. Unfortunately, the appropriate research is not being done. It is interesting to note that, in the New Forest, the most important pasture-woodland in Britain, natural regeneration in most areas is good, or sufficient to maintain the structure of the area, despite grazing by deer, ponies and domestic stock and recreational pressures (Tubbs 1968; Flower 1977).

Complete change of use, particularly to arable farming, the removal of over-mature trees and dead wood, the use of agricultural herbicides and fertilizers, and atmospheric pollution are individually, or collectively, threatening the survival of the flora and fauna of many important pasture-woodlands. Only by adequate legislation and funding for wildlife conservation can these processes be controlled and the loss of the habitat type reduced. Unfortunately, due to past lack of interest, or the lack of awareness of the importance of pasture-woodlands, and to present inadequate funding for wildlife conservation, it seems probable that the present modification and loss of sites will continue.

ACKNOWLEDGEMENTS

Part of this work (PTH) was commissioned by the Nature Conservancy Council in its programme of research into nature conservation. He is grateful to Dr G F Peterken and Dr R C Welch for help and guidance during the course of the survey work. The Nature Conservancy Council and the World Wildlife Fund gave financial help to the surveys of epiphytes (FR). It is a pleasure to acknowledge the landowners, too numerous to name individually, who kindly allowed access to their properties. We are grateful to all those who assisted in the epiphyte survey work, both in the field and with subsequent identification, especially Mr R H Bailey, Dr H J M Bowen, Dr B J Coppins, Mr S Davey, Mr I Day, Mr V Giarvarini, Dr O L Gilbert, Professor D L Hawksworth, Mr P W James, Mr R Jarman, Mr P Lambley, Dr A Pentecost, Dr C Pope, Dr T D V Swinscow, Mrs P Wolseley and Mr R G Woods; and to the many entomologists who contributed information about localities and their faunas, especially Messrs A A Allen, J Cooter, P M Hammond, F A Hunter, C Johnson, P Skidmore, Dr P Hyman and Professor J A Owen. We are also grateful to Dr J P Dempster, Mr J N R Jeffers, Dr M G Morris, Dr O Rackham, Mr R C Steele, Mr A E Stubbs and Dr R C Welch for valuable comments on the original report and, in some cases, on subsequent drafts of this publication. We are very grateful to Mrs S M Weller for typing earlier drafts and to Mrs T E Couch and Miss J M Abblitt for typing the present version.

REFERENCES

Allen, A.A. 1952. Coleoptera at Knole Park, Sevenoaks, Kent. *Entomologist's Rec. J. Var.,* **64**, 224-228.

Allen, A.A. 1956. Maple confirmed as the host-tree of *Gastrallus immarginatus* (Col., Anobiidae) at Windsor. *Entomologist's mon. Mag.,* **92**, 42.

Allen, A.A. 1966a. The rarer Sternoxia (Col.) of Windsor Forest. *Entomologist's Rec. J. Var.,* **78**, 14-23.

Allen, A.A. 1966b. A note on *Hylecoetus dermestoides* (L.) (Col., Lymexylidae). *Entomologist's Rec. J. Var.,* **78**, 79-80.

Appleton, D., Dickson, R. & Else, G.R. 1975. *A provisional list of the insects of Harewood Forest.* Southampton. (Unpublished).

Atty, D.B. 1983. *Coleoptera of Gloucestershire.* Cheltenham. (Privately published).

Bazeley, M.L. 1921. The extent of the English forest in the thirteenth century. *Trans. R. hist. Soc.,* 4th ser., **4**, 140-172.

Birks, H.J.B., Deacon, J. & Peglar, S. 1975. Pollen maps for the British Isles 5,000 years ago. *Proc. R. Soc.,* **189B**, 87-105.

Bond, C.J. 1981. Woodstock Park under the Plantagenet Kings: the exploitation and use of wood and timber in a medieval deer park. *Arboric. J.,* **5**, 201-213.

Brandon, P.F. 1963. *The common lands and wastes of Sussex.* Ph.D. thesis, University of London. (Unpublished).

Buck, F.D. 1955. A provisional list of the Coleoptera of Epping Forest. *Entomologist's mon. Mag.,* **91**, 174-192.

Buckland, P.C. 1975. Synanthropy and the death-watch; a discussion. *Naturalist, Hull,* **100**, 37-42.

Buckland, P.C. 1979. *Thorne Moors: a palaeoecological study of a Bronze Age site.* (Occasional publication no. 8). Birmingham: University of Birmingham, Department of Geography.

Buckland, P.C. & Kenward, H.K. 1973. Thorne Moor: a paleo-ecological study of a Bronze Age site. *Nature, Lond.,* **241**, 405-406

Cantor, L.M. & Hatherley, J. 1979. The medieval parks of England. *Geography,* **64**, 71-85.

Carr, J.W. 1916. *The invertebrate fauna of Nottinghamshire.* Nottingham: Bell.

Carr, J.W. 1935. *The invertebrate fauna of Nottinghamshire, first supplement.* Nottingham: Bell.

Coles, J.M. & Orme, B.J. 1976. The Sweet Track, railway site. *Somerset Levels Pap.,* **2**, 34-65.

Coles, J.M. & Orme, B.J. 1977. Neolithic hurdles from Walton Heath, Somerset. *Somerset Levels Pap.,* **3**, 6-29.

Coope, G.R. 1970. Interpretation of Quaternary insect fossils. *A. Rev. Ent.,* **15**, 97-120.

Cooter, J. 1980. A note on *Ernoporus caucasicus* Lind. (Col., Scolytidae) in Britain. *Entomologist's mon. Mag.,* **116**, 112.

Cooter, J. & Welch, R.C. 1981. *The Coleoptera of Moccas Park, Herefordshire. Appendix I, A list of the Coleoptera known from Moccas Park, Herefordshire.* Abbots Ripton: Institute of Terrestrial Ecology. (Unpublished).

Coppins, B.J. & James, P.W. In press. New or interesting British lichens. *Lichenologist.*

Corke, D., ed. 1978. *Epping Forest - the natural aspect? Essex Nat.,* **2**, 1-79.

Coy, J. 1982. Woodland mammals in Wessex - the archaeological evidence. In: *Archaeological aspects of woodland ecology,* edited by M. Bell & S. Limbrey, 287-296. (BAR International Series 146). Oxford: British Archaeological Reports.

Crowson, R.A. 1962. Observations on Coleoptera in Scottish oakwoods. *Glasg. Nat.,* **18**, 177-195.

Crowson, R.A. 1964. Additional records of Coleoptera from Scottish oakwood sites. *Glasg. Nat.,* **18**, 371-375.

Crowson, R.A. & Hunter, F.A. 1964. Some Coleoptera associated with old trees in Grimsthorpe Park, Lincs. *Entomologist's mon. Mag.,* **100**, 198-200.

Darby, H.C., ed. 1952-77. *The Domesday geography of England.* 6 vols. Cambridge: Cambridge University Press.

Denman, D.R., Roberts, R.A. & Smith, H.J.F. 1967. *Commons and village greens.* London: Leonard Hill.

Donisthorpe, H. St J.K. 1913. The myrmecophilous Coleoptera of Great Britain. In: *The Coleoptera of the British Isles, vol. 6,* by W.W. Fowler & H. St J.K. Donisthorpe, 320-330. London: Reeve.

Donisthorpe, H. St J.K. 1939. *A preliminary list of the Coleoptera of Windsor Forest.* London: Nathaniel Lloyd.

Driscole, D.L. 1977. *The Coleoptera, Hemiptera : Heteroptera and Lepidoptera of Darenth Wood, Kent.* Wye: Nature Conservancy Council. (Unpublished).

Duffy, E.A.J. 1952. Coleoptera: Cerambycidae. *Handbk Ident. Br. Insects,* **5** (12), 1-18.

Edlin, H.L. 1971. Woodland notebook: goodbye to the pollards. *Q. Jl For.,* **65**, 157-165.

Elton, C.S. 1966. *The pattern of animal communities.* London: Methuen.

Fairbairn, W.A. 1972. Dalkeith oakwood. *Scott. For.,* **26**, 5-28.

Ferry, B.W., Baddeley, M.S. & Hawksworth, D.L., eds. 1973. *Air pollution and lichens.* London: Athlone Press.

Fletcher, A. et al. 1982. *Survey and assessment of epiphytic lichen habitats.* 2 vols. (CST report no. 384). Banbury: Nature Conservancy Council.

Flower, N. 1977. *An historical and ecological study of enclosed and unenclosed woods in the New Forest, Hampshire.* Ph.D. thesis, University of London. (Unpublished).

Flower, N. 1980. The management history and structure of unenclosed woods in the New Forest, Hampshire. *J. Biogeogr.,* **7**, 311-328.

Fowler, W.W. 1887-91. *The Coleoptera of the British Isles, vols 1-5.* London: Reeve.

Gardner, A.E., Williams, S.A. *et al.* No date. A manuscript list of the Coleoptera of the New Forest, in the possession of Dr R.C. Welch, Institute of Terrestrial Ecology, Monks Wood Experimental Station, Abbots Ripton.

Gardiner, A.S. 1974. A history of the taxonomy and distribution of the native oak species. In: *The British oak, its history and natural history,* edited by M.G. Morris & F.H. Perring, 13-26. Faringdon: Classey.

Garland, S.P. 1983. Beetles as primary woodland indicators. *Sorby Rec.,* **21**, 3-38.

Gilbert, O.L. 1971. Some indirect effects of air pollution on bark-living invertebrates. *J. appl. Ecol.,* **8**, 77-85.

Girling, M. 1976. Fossil Coleoptera from the Somerset Levels: the Abbot's Way. *Somerset Levels Pap.,* **2**, 28-33.

Girling, M. 1979. Fossil insects from the Sweet Track. *Somerset Levels Pap.,* **5**, 84-93.

Girling, M.A. 1982. Fossil insect faunas from forest sites. In: *Archaeological aspects of woodland ecology,* edited by M. Bell & S. Limbrey, 129-146. (BAR international series 146). Oxford: British Archaeological Reports.

Girling, M. & Greig, J. 1977. Palaeoecological investigations of a site at Hampstead Heath, London. *Nature, Lond.,* **268**, 45-47.

Godwin, H.E. 1975. *The history of the British flora: a factual basis for phytogeography.* 2nd ed. Cambridge: Cambridge University Press.

Grigson, C. 1978. The Late Glacial and Early Flandrian ungulates of England and Wales - an interim review. In: *The effect of man on the landscape: the lowland zone,* edited by S. Limbrey & J.G. Evans, 46-56. (Council for British Archaeology research report no. 21). London: Council for British Archaeology.

Gruffydd, J. St B. 1977. *Protecting historic landscapes.* London: Landscape Institute.

Hadfield, M. 1967. *Landscape with trees.* London: Country Life.

Haines, F.H. 1929. The New Forest as a nature reserve. *Trans. ent. Soc. S. Engl.,* **5**, 76-85.

Hammond, P.M. 1974. Changes in the British coleopterous fauna. In: *The changing flora and fauna of Britain,* edited by D.L. Hawksworth, 323-369. London: Academic Press.

Hammond, P.M. 1979. Beetles in Epping Forest. In: *The wildlife of Epping Forest,* edited by D. Corke. *Essex Nat.,* **4**, 43-60.

Hanson, M.W. 1983. Lords Bushes - the history and ecology of an Epping Forest woodland. *Essex Nat.,* **7**, 1-69.

Harding, P.T. 1976. *The invertebrate fauna of the mature timber habitat: an inventory and survey of areas of conservation value.* (CST report no. 65). Banbury: Nature Conservancy Council.

Harding, P.T. 1977a. *The fauna of the mature timber habitat: second report.* (CST report no. 103). Banbury: Nature Conservancy Council.

Harding, P.T. 1977b. *Moccas Deer Park, Hereford and Worcester, a report on the history, structure and natural history.* Natural Environment Research Council contract report to the Nature Conservancy Council. Abbots Ripton: Institute of Terrestrial Ecology. (Unpublished).

Harding 1978a. *The fauna of the mature timber habitat: third report.* (CST report no. 159). Banbury: Nature Conservancy Council.

Harding, P.T. 1978b. *The invertebrate fauna of the mature timber habitat: survey of areas - site reports : England.* (CST report no. 162). Banbury: Nature Conservancy Council.

Harding, P.T. 1978c. *The invertebrate fauna of the mature timber habitat: survey of areas - site reports: Scotland.* (CST report no. 163). Banbury: Nature Conservancy Council.

Harding, P.T. 1978d. *The invertebrate fauna of the mature timber habitat: survey of areas - site reports: Wales.* (CST report no. 164). Banbury: Nature Conservancy Council.

Harding, P.T. 1978e. *An inventory of areas of conservation value for the invertebrate fauna of the mature timber habitat.* (CST report no. 160). Banbury: Nature Conservancy Council.

Harding, P.T. 1978f. *A bibliography of the occurrence of certain woodland Coleoptera in Britain: with special reference to timber-utilising species associated with old trees in pasture-woodlands.* (CST report no. 161). Banbury: Nature Conservancy Council.

Harding, P.T. 1979a. *Brampton Bryan Park, Hereford & Worcester. A description and survey, with proposals for management.* (CST report no. 262). Banbury: Nature Conservancy Council.

Harding, P.T. 1979b. *Dynevor Deer Park, Dyfed. A description and survey, with proposals for management.* (CST report no. 264). Banbury: Nature Conservancy Council.

Harding, P.T. 1979c. *A survey of the trees at Dunham Massey Park, Greater Manchester.* (CST report no. 263). Banbury: Nature Conservancy Council.

Harding, P.T. 1980. Shute Deer Park, Devon. A pasture-woodland of importance for the conservation of wildlife. *Nat. Devon,* **1**, 71-77.

Harding, P.T. 1981. The conservation of pasture-woodlands. In: *Forest and woodland ecology,* edited by A.S. Gardiner & F.T. Last, 45-48. Cambridge: Institute of Terrestrial Ecology.

Harding, P.T. 1982. A further note on *Ernoporus caucasicus* Lind. (Col., Scolytidae) in Britain. *Entomologist's mon. Mag.,* **118**, 166.

Hawksworth, D.L., Coppins, B.J. & Rose, F. 1974. Changes in the British lichen flora. In: *The changing flora and fauna of Britain,* edited by D.L. Hawksworth, 47-78. London: Academic Press.

Hawksworth, D.L., James, P.W. & Coppins, B.J. 1980. Checklist of British lichen-forming, lichenicolous and allied fungi. *Lichenologist,* **12**, 1-115.

Hawksworth, D.L, & Rose, F.1969. A note on the lichens and bryophytes of the Wyre Forest. *Proc. Birmingham nat. Hist. phil. Soc.,* **21**, 191-197.

Hawksworth, D.L. & Rose, F. 1970. Qualitative scale for estimating sulphur dioxide air pollution in England and Wales using epiphytic lichens. *Nature, Lond.,* **227**, 145-148.

Hawksworth, D.L. & Rose, F. 1976. *Lichens as pollution monitors.* (Institute of Biology studies in biology no. 66). London: Edward Arnold.

Hickin, N.E. 1971. *The natural history of an English forest - the wild life of Wyre.* London: Hutchinson.

Hodge, P. 1980. Coleoptera in Arundel Park (Sussex) during 1979. *Entomologist's mon. Mag.* **115** (1979), 218.

Hunter, F.A. 1977. Ecology of pinewood beetles. In: *Native pinewoods of Scotland,* edited by R.G.H. Bunce & J.N.R. Jeffers, 42-55. Cambridge: Institute of Terrestrial Ecology.

Hunter, F.A. & Johnson, C. 1966. Further notes on Coleoptera associated with old trees in Grimsthorpe Park, Lincs. *Entomologist's mon. Mag.,* **102**, 284.

Jacobi, R.M. 1978. Population and landscape in Mesolithic lowland Britain. In: *The effect of man on the landscape: the lowland zone,* edited by S. Limbrey & J.G. Evans, 75-85. (Council for British Archaeology research report no. 21). London: Council for British Archaeology.

James, P.W., Hawksworth, D.L. & Rose, F. 1977. Lichen communities in the British Isles: a preliminary conspectus. In: *Lichen ecology,* edited by M.R.D. Seaward, 295-413. London: Academic Press.

Johnson, C., Robinson, N.A. & Stubbs, A.E. 1977. *Dunham Park: a conservation report on a parkland of high entomological interest.* (Chief Scientist's Team notes no. 5). London: Nature Conservancy Council.

Kelner-Pillault, J. 1974. Etude écologique du peuplement entomologique des terreaux d'arbres creux (Chataigniers et Saules). *Bull. Ecol.,* **5**, 123-156.

Kennedy, C.E.J. & Southwood, T.R.E. 1984. The number of species of insects associated with British trees: a re-analysis. *J. Anim. Ecol.,* **53**, 455-478.

Kenward, H.K. 1976. Reconstructing ancient ecological conditions from insect remains; some problems and an experimental approach. *Ecol. Entomol.,* **1**, 7-17.

Larkin, P.A. & Elbourn, C.A. 1964. Some observations on the fauna of dead wood in live oak trees. *Oikos,* **15**, 79-92.

Leseigneur, L. 1972. Coléoptères Elateridae de la faune de France continentale et de Corse. *Bull. mens. Soc. linn. Lyon,* **41**, (suppl.).

Marshall, J.E. 1978. The larva of *Aulonium trisulcum* (Fourcroy) (Coleoptera: Colydiidae) and its association with elm bark beetles (*Scolytus* spp.). *Entomologist's Gaz.,* **29**, 59-69.

Massee, A.M. 1964. *Some of the more interesting Coleoptera and Hemiptera-Heteroptera recorded at Moccas Deer Park, Herefordshire.* Manuscript list in the possession of the Nature Conservancy Council.

Mitchell, B., Staines, B.W. & Welch, D. 1977. *Ecology of red deer.* Cambridge: Institute of Terrestrial Ecology.

Morris, M.G. 1974. Oak as a habitat for insect life. In: *The British oak, its history and natural history,* edited by M.G. Morris & F.H. Perring, 274-297. Faringdon: Classey.

Osborne, P.J. 1965. The effect of forest clearance on the distribution of the British insect fauna. *Proc. XII Int. Congr. Ent., London,* 1964, 456-457.

Osborne, P.J. 1972. Insect faunas of late Devensian and Flandrian age from Church Stretton, Shropshire. *Phil. Trans. R. Soc,* Ser. B, **263**, 327-367.

Owen, J.A. 1984. *Rare woodland beetles occurring in the Windsor area.* Newbury: Nature Conservancy Council. (Unpublished).

Oxford, P.M. 1975. A late post-glacial vegetation sequence from Epping Forest. *Horizon,* **24**, 85-102.

Palm, T. 1959. Die Holz- und Rinden-Käfer der süd- und mittelschwedischen Laubbaüme. *Opusc. ent.,* Suppl. 16.

Paviour-Smith, K. 1973. Review of Stubbs, A.E. (1972). *Wildlife conservation and dead wood. Entomologist's mon. Mag.,* **109**, 52-53.

Pennington, W. 1969. *The history of British vegetation.* London: English Universities Press.

Peterken, G.F. 1969. Development of vegetation in Staverton Park, Suffolk. *Fld Stud.,* **3**, 1-39.

Peterken, G.F. 1974. Developmental factors in the management of British woodlands. *Q. Jl For.,* **68**, 141-149.

Peterken, G.F. 1977a. Habitat conservation priorities in British and European woodlands. *Biol. Conserv.,* **11**, 223-236.

Peterken, G.F. 1977b. General management principles for nature conservation in British woodlands. *Forestry,* **50**, 27-48.

Peterken, G.F. 1981. *Woodland conservation and management.* London: Chapman and Hall.

Pigott, C.D. & Walters, S.M. 1954. On the interpretation of the discontinuous distributions shown by certain British species of open habitats. *J. Ecol.,* **42**, 95-116.

Pitt, J. 1984. Ancient pollards of Lullingstone Park, Kent. *Trans. Kent Fld Club,* **9**, 129-142.

Pope, R.D. 1977. A check list of British insects. Part 3: Coleoptera and Strepsiptera. *Handbk Ident. Br. Insects,* **11**(3), 1-105.

Prince, H.C. 1958. Parkland in the English landscape. *Amat. Hist.,* **3**, 332-349.

Rackham, O. 1977. *Trees and woodland in the British Landscape.* London: Dent.

Rackham, O. 1977. Neolithic woodland management in the Somerset Levels: Garvin's, Walton Heath, and Rowland's Tracks. *Somerset Levels Pap.,* **3**, 65-71.

Rackham, O. 1980. *Ancient woodland, its history, vegetation and uses in England.* London: Edward Arnold.

Radley, J. 1961. Holly as a winter feed. *Agric. Hist. Rev.,* **9**, 89-92.

Ratcliffe, D.A., ed. 1977. *A nature conservation review.* Cambridge: Cambridge University Press.

Rose, F. 1957. The importance of the study of disjunct distributions to progress in understanding the British flora. In: *Progress in the study of the British flora,* edited by J.E. Lousley, 61-78. London: Botanical Society of the British Isles.

Rose, F. 1974. The epiphytes of oak. In: *The British oak, its history and natural history,* edited by M.G. Morris & F.H. Perring, 250-273. Faringdon: Classey.

Rose, F. 1976. Lichenological indicators of age and environmental continuity in woodlands. In: *Lichenology: progress and problems,* edited by D.H. Brown, D.L. Hawksworth & R.H. Bailey, 279-307. London: Academic Press.

Rose, F. & Harding, P.T. 1978. *Pasture-woodlands in lowland Britain and their importance for the conservation of the epiphytes and invertebrates associated with old trees.* (CST report no. 211). Banbury: Nature Conservancy Council.

Rose, F., Hawksworth, D.L. & Coppins, B.J. 1970. A lichenological excursion through the north of England. *Naturalist, Hull,* **95**, 49-55.

Rose, F. & James, P.W. 1974. Regional studies on the British lichen flora, 1. The corticolous and lignicolous species of the New Forest, Hampshire. *Lichenologist,* **6**, 1-72.

Rose, F. & Wallace, E.C. 1974. Changes in the bryophyte flora of Britain. In: *The changing flora and fauna of Britain,* edited by D. L. Hawksworth, 27-46. London: Academic Press.

Rose, F. & Wolseley, P. 1984. Nettlecombe Park - its history and its epiphytic lichens: an attempt at correlation. *Fld Stud.,* **6**, 117-148.

Shaw, S. 1956. On the occurrence of *Hylecoetus dermestoides* (L.) and *Lymexylon navale* (L.) (Col. Lymexylidae) in Lancashire and Cheshire. *J. Soc. Br. Ent.,* **5**, 172.

Sheail, J. 1976. *Nature in trust: the history of nature conservation in Britain.* London: Blackie.

Shirt, D.B., ed. 1986. *British Red Data Books : 2 Insects.* Peterborough: Nature Conservancy Council.

Smith, A.G. 1970. The influence of Mesolithic and Neolithic man on British vegetation: a discussion. In: *Studies of the vegetational history of the British Isles,* edited by D. Walker & R.G. West, 81-96. Cambridge: Cambridge University Press.

Smith, S.E.A. 1978. *The trees of Dunham Park.* M.Sc. dissertation, University of Salford. (Unpublished).

Southwood, T.R.E. 1961. The number of species of insect associated with various trees. *J. Anim. Ecol.,* **30**, 1-8.

Spray, M. 1981. Holly as a fodder in England. *Agric. Hist. Rev.,* **29**, 97-110.

Streeter, D.T. 1974. Ecological aspects of oak woodland conservation. In: *The British oak, its history and natural history,* edited by M.G. Morris & F.H. Perring, 13-26. Faringdon: Classey.

Stubbs, A.E. 1972. Wildlife conservation and dead wood. *J. Devon Trust Nature Conserv.,* suppl., 1-18.

Tittensor, A. & Tittensor, R. 1977. *Natural history of The Mens, Sussex.* Horsham: Horsham Natural History Society.

Tittensor, R.M. 1978. A history of The Mens: a Sussex woodland common. *Sussex Archaeol. Collect.,* **116**, 347-374.

Tittensor, R.M. 1981. A sideways look at nature conservation in Britain. *Discuss. Pap. Conserv.,* no. 29.

Tubbs, C.R. 1968. *The New Forest: an ecological history.* Newton Abbot: David & Charles.

Turner, D. & Dillwyn, L.W. 1805. *The botanist's guide through England and Wales.* 2 vols. London: Phillips & Fardon.

Walker, J.J. 1890. Coleoptera at Cobham Park, Kent. *Entomologist's mon. Mag.,* **26**, 9-11.

Welch, R.C., ed. 1972. *Windsor Forest study.* Abbots Ripton: Nature Conservancy. (Unpublished).

Welch, R.C. 1973. Coleoptera. In: *Monks Wood, a nature reserve record,* edited by R.C. Steele & R.C. Welch, 212-233. Huntingdon: Nature Conservancy.

Welch, R.C. 1975. Coleoptera. In: *Bedford Purlieus, its history, ecology and management,* edited by G.F. Peterken & R.C. Welch,159-186. (Monks Wood Symposium no. 7). Abbots Ripton: Institute of Terrestrial Ecology.

Welch, R.C. 1977. *The Coleoptera of Grimsthorpe Park, Lincolnshire.* Abbots Ripton: Institute of Terrestrial Ecology. (Unpublished).

Welch, R.C. 1968-80. The Coleoptera of Monks Wood National Nature Reserve, Huntingdonshire. Supplements 1-5. *Entomologist's Gaz.,* **19**, 9-20; **21**, 133-141; **22**, 235-240; **26**, 119-126; **31**, 263-273.

Welch, R.C. & Cooter, J. 1981. *The Coleoptera of Moccas Park, Herefordshire.* Abbots Ripton: Institute of Terrestrial Ecology. (Unpublished).

Welch, R.C. & Harding, P.T. 1974. A preliminary list of the fauna of Staverton Park, Suffolk. Part 2, Insecta: Coleoptera. *Suffolk natur. hist.,* **16**, 287-304.

West, R.G. 1968. *Pleistocene geology and biology.* London: Longman.

APPENDIX 1

SURVEY REPORTS ON LICHENS IN LOWLAND PASTURE-WOODLANDS

The following list of mainly unpublished reports, drawn from Fletcher *et al.* (1982), includes the main survey reports on lichens in pasture-woodlands in lowland Britain. Most reports are by F Rose, but a few are by other members of the British Lichen Society. Most of the reports are deposited at the appropriate regional offices of the Nature Conservancy Council.

The reports are listed separately for England, Scotland and Wales, with those for England being listed under the appropriate counties or regions.

ENGLAND

BERKSHIRE

1. Rose, F. & Showell, J. 1968. Vice County 22. Windsor Forest 17/2/68. (3 pp).

BUCKINGHAMSHIRE

2. Anon. Undated. Burnham Beeches lichens. (det. Rose, F., species list, 1 p).

3. Rose, F. 1977. Chequers Park. (MS 1 p).

CORNWALL

4. Lambley, P.W. 1976. Report on fieldwork on lichens carried out in west Cornwall with the aid of a grant from the World Wildlife Fund. (10 pp plus maps and species lists).

5. Rose, F. Undated. Report on surveys of woodland and parkland carried out in Cornwall, 1-4 October 1971. (6 pp).

CUMBRIA

6. Rose, F. 1971. A survey of the woodlands of the Lake District, and an assessment of their conservation value based upon structure, age of trees and lichen and bryophyte epiphyte flora. (31 pp).

7. Rose, F. 11.9.1973. Interim report on surveys done in Westmorland and Cumberland, August 29-September 1 1973. (6 pp).

8. Rose, F. Undated. Report on woodland sites surveyed in Cumbria, September 16-19 1975. (5 pp).

DEVON

9. Hawksworth, D.L. 9.3.1977. Lichen Survey of Selected Sites in North Devon. (31 pp, including maps and species lists).

DORSET

10. Rose, F. 8.5.1975. Report on survey of Tyneham Ranges for bryophytes and lichens. 18-29 April 1975. (5 pp, plus MS).

DURHAM

(See number 43)

EAST ANGLIA (Essex, Norfolk, Suffolk)

11. Davis, B.N.K. & Lambley, P.W. 1972. Sotterley Park, Suffolk. (3 pp, plus map).

12. Peterken, G.F. & Rose, F. 10.7.1970. Felbrigg Wood, Norfolk. (5 pp).

13. Rose, F. Undated. Corticolous and lignicolous lichens of Staverton Park. (4 pp, plus species list).

14. Rose, F. 26.3.1969. Preliminary report on deciduous woodlands and old parklands containing fragments of old deciduous woodland, in Norfolk and Suffolk. (I Suffolk 3 pp; II Norfolk 17 pp).

15. Rose, F. 10.10.1971. Report on Sotterley Park, East Suffolk. (2 pp).

16. Rose, F. 23.10.1971. Report on sites visited in East Anglia, June 1970, March 1971, July 1971. (23 pp, plus maps and species lists).

17. Rose, F. 10.5.1972. Hintlesham Great Wood. (1 p).

18. Rose, F. 10.5.1972. Wayland Wood. (1 p).

19. Rose, F. 10.5.1972. Hull Wood, Glandford. (1 p).

20. Rose, F. 10.5.1972. Sheringham Park. (1 p).

21. Rose. F. 25.4.1973. Report on survey work on cryptogamic vegetation in Essex and Suffolk, 16-19 March 1973. (3 pp).

22. Rose, F. 7.1974. Report on work done in East Anglia in botanical survey, 16-19 July 1974. (18 pp, plus 1 page supplement).

GLOUCESTERSHIRE

23. Rose, F. 4.9.1968. Report on survey carried out in the Forest of Dean, April and May 1968. (11 pp).

HEREFORD/WORCESTER

24. Rose, F. 1.10.1968. Great Wood, Vowchurch, Herefordshire. (1 p).

25. Rose, F. 1.10.1968. Moccas Park. (1 p).

26. Rose, F. 4.10.1968. Banses Wood, south-east of Cockyard, Herefordshire, SO413334, May 31 1968. (1 p).

27. Rose, F. 1972. Eastnor Park, Herefordshire, Holme Lacy Park and Wood, Kentchurch Park, Homend Park. (4 pp).

28. Rose, F. 1972. Brampton Bryan Park, Herefordshire. (3 pp, including species list and map).

29. Rose, F. 1972. Downton Castle and Gorge, Herefordshire. (3 pp, including species list).

30. Rose, F. 1974. Survey of woodland and parkland sites in the south-west midlands. (8 pp).

31. Rose, F. 8.8.1975. The lichens and bryophytes of sites in Hereford, Salop and Staffordshire based on a survey made from April 21-24 1975. (10 pp).

32. Rose, F. 1976. Wye Gorge - below Symonds Yat, Herefordshire. (3 pp).

KENT

33. Anon. (Rose, F) Undated. Survey report on woodland and forest relict areas of the Wealden region. (27 pp).

LEICESTERSHIRE

34. Fletcher, A. 1980. Report on lichens of Stanford Park, Leicestershire. (MS 5 pp, including map).

35. Hawksworth, D.L. 10.1972. Leicestershire & Rutland. Report on the lichen flora and vegetation of sites of scientific importance for lichens together with notes on other notified NNR and SSSI localities. (10 pp).

OXFORDSHIRE

36. Bowen, H.J.M., Rose, F. & Mason, J. 1972. List of species from Wychwood, Oxon, seen in April 1972. (5 pp plus map).

37. Rose, F. 1977. Stonor Park. (MS 1 p).

SHROPSHIRE

38. Rose, F. 24.4.1969. Wyre Forest (visited 28 and 29 May 1968). (21 pp).

(See also number 31)

STAFFORDSHIRE

(See number 31)

WILTSHIRE

39. Rose, F. Undated. Cranborne Chase-Rushmore Park area. (2 pp, species list).

40. Rose, F. 1975. Report on the conservation of epiphytic lichens and bryophytes on the Cranborne Chase Estate, Wiltshire and Dorset. Surveyed August 28 1975. (MS 5 pp).

41. Rose, F. 1976. Report on survey of cryptogam flora (bryophytes and lichens) of the Longleat Estate. Surveyed 8.1969, 11.1973, 27.11.1974. (4 pp, plus supplement, undated 2 pp, species list).

42. Rose, F. 1976. Report on survey of Cranborne Chase 23.3.1976 and 24.3.1976. (5 pp plus maps).

YORKSHIRE

43. Rose, F. 1971. Survey report on woodland sites in North Yorkshire and Co Durham. (17 pp).

SCOTLAND

44. Coppins, B.J. 1977. Field meeting at Girvan, Ayrshire. *Lichenologist,* **9**, 153-157.

45. Coppins, B.J. 3.1977. Lichen flora of Dalkeith Old Wood. (1 p, including species list).

46. Rose, F. & James, P.W. 1976. A survey of the lichens of Galloway and western Dumfries-shire. (MS 25 pp).

WALES

47. Anon - NCC Undated. Great Wood, Gregynog SSSI. (1 p, including map).

48. Rose, F. 24.7.1971. Report on the cryptogamic vegetation of woodland areas in north Wales. (9 pp).

49. Rose, F. 2.1.1973. Interim report on survey work in South Wales, June 6-9 1972. (8 pp).

50. Rose, F. 2.1.1973. Report on woodland surveys in South Wales. June and September 1972. (9 pp).

51. Rose, F. 10.5.1975. Report on sites surveyed in the (former) Merionethshire, March 27-April 2 1975. (MS 6 pp).

APPENDIX 2

A LIST OF THE SAPROXYLIC COLEOPTERA OF PASTURE-WOODLANDS

An earlier attempt (Harding 1977a) to list and grade species of invertebrates as indicators of the continuity of dead-wood habitats in ancient woodlands, particularly pasture-woodlands, was only partly successful. The original list of Coleoptera was compiled in association with 4 coleopterists, A A Allen, F A Hunter, C Johnson and P Skidmore, who helped achieve a fairly well-balanced list of species with limited regional bias. Subsequently, Harding (1978f) examined the occurrence of 99 species of Coleoptera (listed as Grades 1 and 2 in the earlier list) and demonstrated that many species are known almost exclusively from areas of ancient pasture-woodland. However, some species in the original list are far too widespread to be considered reliable indicators of habitat continuity.

The following list of 196 mainly saproxylic Coleoptera is derived from the earlier list (Harding 1977a), but has been modified in several ways. Several species have been added to, or removed from, the original list, mainly in response to information supplied by C Johnson and P M Hammond and published information in Hammond (1979) and Garland (1983). Many additional records have been examined, mainly from the period 1978-84, which have caused the grouping of species in the list to be re-appraised.

The list remains tentative until the ecology and distribution of many species are better understood. Use of the list to evaluate and compare sites must be made with its limitations in mind. It is a list of mainly saproxylic species believed to be associated with dead-wood habitats in pasture-woodlands and is not, therefore, a comprehensive list of woodland indicator species. It is a national list for pasture-woodlands in lowland Britain, in which regional variations can be accommodated only to a limited extent. For example, several species listed in Group 3 are probably reliable indicators of habitat continuity at the limits of their range in Britain, but can occur in a wider range of habitats (eg hedges and gardens) at the centre of their range.

The species have been grouped (Groups 1-3) according to the extent to which they have been consistently recorded from areas of ancient woodland with continuity of dead-wood habitats, particularly in pasture-woodlands. Most of the species listed occur, at some part of their life cycle, closely associated with the bark, sapwood, heartwood or decayed wood of broadleaved trees or shrubs, or with xylophilous fungi.

Group 1 : Species which are known to have occurred in recent times only in areas believed to be ancient woodland, mainly pasture-woodland.

Group 2 : Species which occur mainly in areas believed to be ancient woodland with abundant dead-wood habitats, but which also appear to have been recorded from areas that may not be ancient or for which the locality data are imprecise.

Group 3 : Species which occur widely in wooded land, but which are collectively characteristic of ancient woodland with dead-wood habitats.

Notes in table

(a) Insufficient information is available about the present distribution of these species to be certain that they belong in this group.

(b) Occasionally imported in timber, etc.

(c) A 'leaf beetle' not associated with timber or dead wood, but known to occur at the New Forest, Windsor Forest and Sherwood Forest, mainly on very old oaks.

For species included in the *British Red Data Book* for insects (Shirt 1986), the *Red Data Book* threat category is included.

Nomenclature follows Pope (1977) and subsequent corrections and additions published in *Antenna.*

		Saproxylic fauna group	Red Data Book category
Carabidae:	*Calosoma inquisitor*	3	
Histeridae:	*Plegaderus dissectus*	2	
	Abraeus granulum	1	
	Aeletes atomarius	1	3
Ptiliidae:	*Ptenidium gressneri*	2	3
	P. turgidum	2	
	Micridium halidaii	1	1
	Ptinella limbata	2	1
Scydmaenidae:	*Eutheia formicetorum*	1	2
	E. linearis	1	1
	Stenichnus bicolor	3	
	S. godarti	1	
	Microscydmus minimus	1	2
	Euconnus pragensis	1	1
	Scydmaenus rufus	3	
Staphylinidae:	*Phyllodrepa nigra*	1	2
	Xantholinus angularis	3	
	Velleius dilatatus	1	1
	Quedius aetolicus	3	
	Q. maurus	3	
	Q. microps	3	
	Q. scitus	3	
	Q. ventralis	3	
	Q. xanthopus	3	
	Euryusa optabilis	2	2
	E. sinuata	2	1
	Tachyusida gracilis	1	1
Pselaphidae:	*Bibloporus minutus*	2	3
	Euplectus brunneus	1(a)	3
	E. nanus	1	
	E. punctatus	1	
	Plectophloeus nitidus	1	1
	Trichonyx sulcicollis	2	3
	Batrisodes buqueti	1	1
	B. delaporti	1	1
	B. venustus	2	3

		Saproxylic fauna group	Red Data Book category
Lucanidae:	*Sinodendron cylindricum*	3	
Scarabaeidae:	*Gnorimus variabilis*	1	1
Scirtidae:	*Prionocyphon serricornis*	2	3
Buprestidae:	*Agrilus pannonicus*	2(b)	2
Elateridae:	*Lacon querceus*	1	1
	Ampedus cardinalis	1	2
	A. cinnabarinus	1	3
	A. elongatulus	3	
	A. nigerrimus	1	1
	A. pomonae	1	
	A. pomorum	3	
	A. ruficeps	1	1
	A. rufipennis	1	2
	Ischnodes sanguinicollis	2	
	Procraerus tibialis	1	2
	Megapenthes lugens	1	1
	Limoniscus violaceus	1	1
	Stenagostus villosus	3	
	Selatosomus bipustulatus	3	
	Elater ferrugineus	1	1
Throscidae:	*Trixagus brevicollis*	1	3
Eucnemidae:	*Eucnemis capucina*	1	1
	Dirhagus pygmaeus	3	3
	Melasis buprestoides	3	
Cantharidae:	*Malthodes brevicollis*	1	3
	M. crassicornis	2	3
Lycidae:	*Pyropterus nigroruber*	2	3
	Platycis cosnardi	1	1
	P. minutus	3	
Dermestidae:	*Globicornis nigripes*	2	1
	Ctesias serra	3	
	Trinodes hirtus	1	3
Anobiidae:	*Xestobium rufovillosum*	3	
	Gastrallus immarginatus	1	1
	Xyletinus longitarsis	3	

		Saxproxylic fauna group	Red Data Book category
	Dorcatoma chrysomelina	2	
	D. dresdensis	2	1
	D. flavicornis	3	
	D. serra	2	
	Anitys rubens	1	
Ptinidae:	*Ptinus palliatus*	3	
	P. subpilosus	2	
Lyctidae:	*Lyctus brunneus*	3	
Phloiophilidae:	*Phloiophilus edwardsi*	3	
Peltidae:	*Thymalus limbatus*	3	
Cleridae:	*Tillus elongatus*	3	
	Opilo mollis	3	
	Thanasimus formicarius	3	
	Korynetes caeruleus	3	
Melyridae:	*Aplocnemus nigricornis*	3	
	A. pini	3	
	Hypebaeus flavipes	1	1
Lymexylidae:	*Hylecoetus dermestoides*	3	
	Lymexylon navale	1	2
Nitidulidae:	*Carpophilus sexpustulatus*	3	
	Epuraea angustula	3	
Rhizophagidae:	*Rhizophagus nitidulus*	3	
	R. oblongicollis	1	1
Cucujidae:	*Uleiota planata*	1(b)	2
	Pediacus depressus	2	
	P. dermestoides	3	
	Laemophloeus monilis	1(a)	1
	Notolaemus unifasciatus	2	3
Sylvanidae:	*Silvanus bidentatus*	2	3
	S. unidentatus	3	
Cryptophagidae:	*Cryptophagus micaceus*	1	3
	Atomaria lohsei	1	3
Biphyllidae:	*Biphyllus lunatus*	3	
	Diplocoelus fagi	2	

		Saxproxylic fauna group	Red Data Book category
Erotylidae:	*Triplax aenea*	3	
	T. lacordairii	3	3
	T. russica	3	
	T. scutellaris	3	3
	Tritoma bipustulata	3	
Cerylonidae:	*Cerylon fagi*	3	
Endomychidae:	*Symbiotes latus*	3	
Lathridiidae:	*Lathridius consimilis*	1	
	Enicmus brevicornis	2	
	E. rugosus	2	2
	Dienerella separanda	2	
	Corticaria alleni	1	
	C. fagi	1	2
	C. longicollis	1	
Cisidae:	*Cis coluber*	2	3
Mycetophagidae:	*Pseudotriphyllus suturalis*	3	
	Triphyllus bicolor	3	
	Mycetophagus atomarius	3	
	M. piceus	3	
Colydiidae:	*Synchita humeralis*	3	
	S. separanda	1	3
	Cicones variegata	2	
	Bitoma crenata	3	
	Colydium elongatum	1	3
	Teredus cylindricus	1	1
	Oxylaemus cylindricus	2(a)	App
	O. variolosus	2(a)	3
Tenebrionidae:	*Eledona agricola*	3	
	Corticeus unicolor	2	3
	Prionychus ater	3	
	P. melanarius	1	2
	Pseudocistela ceramboides	2	
	Mycetochara humeralis	3	
Tetratomidae:	*Tetratoma ancora*	3	
	T. desmaresti	3	
	T. fungorum	3	

		Saproxylic fauna group	Red Data Book category
Pyrochroidae:	*Pyrochroa coccinea*	3	
Melandryidae:	*Hallomenus binotatus*	3	
	Orchesia undulata	3	
	Anisoxya fuscula	3	3
	Abdera biflexuosa	3	
	A. quadrifasciata	1	
	Phloiotrya vaudoueri	2	
	Hypulus quercinus	2	2
	Melandrya barbata	1	1
	M. caraboides	3	
	Conopalpus testaceus	3	
Scraptiidae:	*Scraptia dubia*	1(a)	
	S. fuscula	1	
	S. testacea	1	
	Anaspis schilskyana	1	1
Mordellidae:	*Tomoxia biguttata*	1	3
	Mordella aculeata	3	
	M. villosa	3	
Oedemeridae:	*Ischnomera caerulea*	3	
	I. cinerascens	1	2
	I. sanguinicollis	1	
Aderidae:	*Aderus brevicornis*	1(a)	3
	A. oculatus	3	
Cerambycidae:	*Prionus coriarius*	3	
	Grammoptera ustulata	1	3
	G. variegata	3	
	Leptura scutellata	1	
	Strangalia aurulenta	3	
	S. quadrifasciata	3	
	S. revestita	2	3
	Pyrrhidium sanguineum	1	1
	Phymatodes testaceus	3	
	Mesosa nebulosa	2	3
	Saperda scalaris	3	
Chrysomelidae:	*Cryptocephalus querceti*	1(c)	2

		Saproxylic fauna group	Red Data Book category
Anthribidae:	*Platyrhinus resinosus*	3	
	Tropideres niveirostris	3	3
	T. sepicola	1	3
	Platystomos albinus	3	
Curculionidae:	*Pentarthrum huttoni*	3	
	Mesites tardii	3	
	Cossonus parallelepipedus	3	
	Phloeophagus truncorum	1	
	Dryophthorus corticalis	1	1
	Trachodes hispidus	3	
Scolytidae:	*Xyloterus domesticus*	3	
	X. lineatus	3	
	X. signatus	3	3
	Ernoporus caucasicus	1	1
	E. fagi	3	
	Xyleborus dispar	3	3
	X. dryographus	3	
	X. saxeseni	3	
Platypodidae:	*Platypus cylindrus*	3	3

APPENDIX 3

THE OCCURRENCE OF THREATENED SAPROXYLIC COLEOPTERA AT SELECTED PASTURE-WOODLANDS

Several authors have listed and compared the coleopterous fauna of important pasture-woodland sites (see page 85). The following table includes 74 species of saproxylic Coleoptera, drawn from Groups 1 and 2 in Appendix 2, and which are listed as threatened in the *British Red Data Book* for insects (Shirt 1986). The occurrence of these species at 9 pasture-woodlands and 2 formerly coppiced woodlands (Darenth Woods and Monks Wood) is compared in the Table. Because faunal lists are available for only very few woodland sites, the comparison is not comprehensive. Scattered records exist for many other pasture-woodland sites, but it would not be meaningful to compare sites with only isolated records with these better documented sites. Many of the species have been recorded from additional sites, but the Table is limited to those sites with at least 5 of these species recorded. The occurrence of these species at some other pasture-woodland sites is summarized in Table 6 (page 50).

The main sources of records are listed below, but a complete list of sources is retained by PTH.

MAIN SOURCES OF RECORDS

Windsor Forest and Windsor Park, Berkshire
    Donisthorpe 1939; Welch 1972; Owen 1984

New Forest, Hampshire
    Gardner *et al.* MS

Sherwood Forest, Nottinghamshire
    Carr 1916, 1935

Moccas Park NNR, Hereford/Worcester
    Massee MS 1964; Harding 1977b; Cooter & Welch MS 1981; Welch & Cooter MS 1981

Epping Forest, Essex
    Buck 1955; Hammond 1979

Richmond Park, Surrey
    Fowler 1887-91; P.M. Hammond & J. Parry pers. comm.

Dunham Park, Greater Manchester
    Johnson *et al.* 1977

Darenth Woods, Kent (mainly 19th century records)
    Fowler 1887-91; Driscole 1977

Arundel Park, West Sussex
    Hodge 1980

Staverton Park, Suffolk
    Welch & Harding 1974

Monks Wood NNR, Cambridgeshire
    Welch 1973; Welch 1968-80

	Red Data Book category	Saproxylic fauna group	Windsor Forest/Park	New Forest	Sherwood Forest	Moccas Park	Epping Forest	Richmond Park	Dunham Park	Darenth Woods	Arundel Park	Staverton Park	Monks Wood
Aderus brevicornis	3	1	+	+									
Aeletes atomarius	3	1	+	+		+							
Agrilus pannonicus	2	2	+	+	+					+			
Ampedus cardinalis	2	1	+		+	+		+					
A. cinnabarinus	3	1	(+)	+	+		+				+		
A. nigerrimus	1	1	+				+						
A. ruficeps	1	1	+										
A. rufipennis	2	1	+			+							
Anaspis schilskyana	[illegible]	1				+							
Atomaria lohsei	3	1		+									
Batrisodes buqueti	1	1	+										
B. delaporti	1	1	+										
B. venustus	3	2	+	+	+	+	+					+	
Bibloporus minutus	3	2	+	+			+		+			+	
Cis coluber	3	2	+	+									
Colydium elongatum	3	1		+									
Corticaria fagi	2	1	+										
Corticeus unicolor	3	2		+	+				+				
Cryptocephalus querceti	2	1	+	+	+								
Cryptophagus micaceus	3	1	+	+				+					
Dorcatoma dresdensis	1	2	+	+									

	Red Data Book category	Saproxylic fauna group	Windsor Forest/Park	New Forest	Sherwood Forest	Moccas Park	Epping Forest	Richmond Park	Dunham Park	Darenth Woods	Arundel Park	Staverton Park	Monks Wood
Dryophthorus corticalis	1	1	+										
Elater ferrugineus	1	1	+					+		+			
Enicmus rugosus	2	2	+		+		+						
Ernoporus caucasicus	1	1				+							
Eucnemis capucina	1	1	+	+									
Euconnus pragensis	1	1	+										
Euryusa optabilis	2	2	+	+			+						
E. sinuata	1	2	+										
Eutheia formicetorum	2	1	+	+									
E. linearis	1	1	+	+	+								
Gastrallus immarginatus	1	1	+										
Globicornis nigripes	1	2	+										
Gnorimus variabilis	1	1	+										
Grammoptera ustulata	3	1	+	+									
Hypebaeus flavipes	1	1				+							
Hypulus quercinus	2	2								+			+
Ischnomera cinerascens	2	1				+							
Lacon querceus	1	1	+										
Laemophloeus monilis	1	1									+		
Limoniscus violaceus	1	1	+										
Lymexylon navale	2	1	+	+		+		+	+				

	Red Data Book category	Saproxylic fauna group	Windsor Forest/Park	New Forest	Sherwood Forest	Moccas Park	Epping Forest	Richmond Park	Dunham Park	Darenth Woods	Arundel Park	Staverton Park	Monks Wood
Malthodes brevicollis	3	1		+		+							
M. crassicornis	3	2	+			+	+					+	
Megapenthes lugens	1	1	+	+			+						
Melandrya barbata	1	1		+						+			
Mesosa nebulosa	3	2	+	+		+							+
Micridium halidaii	1	1	+		+			+					
Microscydmus minimus	2	1	+		+								
Notolaemus unifasciatus	3	2	+	+		+	+	+					
Oxylaemus cylindricus	APP	2		+									
O. variolosus	3	2								+			+
Phyllodrepa nigra	2	1	+	(+)									
Plectophloeus nitidus	1	1	+		+	+							
Prionocyphon serricornis	3	2	+	+	+		+	+	+	+			
Prionychus melanarius	2	1			+						+	+	
Procraerus tibialis	2	1	+	+	+	+		+				+	
Ptenidium gressneri	3	2	+	+	+		+		+				+
Ptinella limbata	1	2		+	+								
Pyropterus nigroruber	3	2			+								
Pyrrhidium sanguineum	1	1				+							
Rhizophagus oblongicollis	1	1	+		+		+	+					
Silvanus bidentatus	3	2	+	+			+		+		+		

	Red Data Book category	Saproxylic fauna group	Windsor Forest/Park	New Forest	Sherwood Forest	Moccas Park	Epping Forest	Richmond Park	Dunham Park	Darenth Woods	Arundel Park	Staverton Park	Monks Wood
Strangalia revestita	3	2	+							+			
Synchita separanda	3	1	+										
Tachyusida gracilis	1	1	+										
Teredus cylindricus	1	1	+		+								
Tomoxia biguttata	3	1	+	+				+			+		
Trichonyx sulcicollis	3	2		+							+		
Trinodes hirtus	3	1	+					+	+				
Trixagus brevicollis	3	1	+			+		+					
Tropideres sepicola	3	1		+		+							+
Uleiota planata	2	1	+					+					
Velleius dilatatus	1	1	+	+		+							
TOTAL (excluding doubtful records)			54	34	19	19	13	13	7	7	6	5	5

(+) Species recorded, but some doubt is attached to the validity of the record by relevant experts.